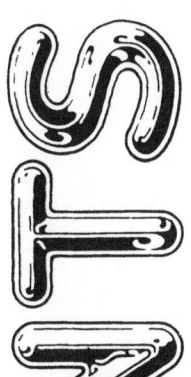
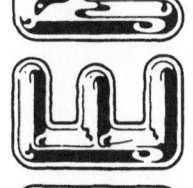

EXPERIMENTS IN MODERN ANALYTICAL CHEMISTRY

W0037529

D Kealey

Senior Lecturer
Department of Chemistry
Kingston Polytechnic

Springer Science+Business Media, LLC

British Library Cataloguing in Publication Data

Kealey, D
 Experiments in modern analytical chemistry.
 1. Chemistry, Analytic   2. Chemistry-Experiments
 I. title
 543'.00724   QD75.2

 ISBN 978-0-216-91802-3   ISBN 978-1-4899-6635-3 (eBook)
          DOI 10.1007/978-1-4899-6635-3

Library of Congress Cataloging-in-Publication Data

Kealey, D. (David)
 Experiments in modern analytical chemistry.

 Includes index.
 1. Chemistry, Analytic--Laboratory manuals.
 I. Title.
 QD76.K43 1985   543'.007'8   85-26669

Preface

The purpose of this book is to provide undergraduate students and teachers of analytical chemistry with a range of experiments that are representative of the techniques and methods currently used in industrial and other analytical laboratories.   It was inevitable that in a book of limited length, several important techniques would have to be omitted. The final choice of experiments to be included was therefore based largely on the responses to a questionnaire recently circulated by the Publishers. From these it became clear that there is considerably more interest in techniques such as titrimetry, gas chromatography, ultraviolet/visible spectrometry, infrared spectrometry and atomic absorption spectrometry than many others and particularly those requiring very costly instrumentation. It is hoped to expand the range of experiments in any future revised edition.   I am grateful to all those who returned a questionnaire and for all the additional helpful comments that were received.

The book has been divided into two sections. The first contains material of a general nature relevant to the practice of analytical chemistry, including a summary of the use of statistical methods for the assessment of quantitative data.

The second section consists of 22 experiments divided into two groups on the basis of the cost/complexity of the apparatus and instrumentation required. Each experiment includes a brief introductory section that summarizes the theoretical basis and some typical applications of the analytical technique(s) involved. Each experiment also includes a set of graded discussion questions, answers to which should form an integral part of a student's experimental report. It is hoped that by varying the degree of difficulty of the questions, both able and less able students will be encouraged to attempt at least some.

The experiments, or modifications of them, have been used at Kingston Polytechnic on a number of undergraduate courses including those leading to LRSC (formerly LRIC), BTEC (formerly HND and HNC), GRSC (formerly GRIC) and BSc, and in suitably extended forms for postgraduate students. Several of the experiments are adaptations of ones originated by colleagues at the Polytechnic, and I am particularly grateful to Mr F.W. Fifield, Mr G.C. Shone and Dr A.K. Slawinski for providing the basic ideas.

D.K.

Contents

EXPERIMENTS

Group A: Experiments requiring little or no specialized equipment or
          instrumentation

Introduction

Modern analytical chemistry employs a range of techniques that vary from simple qualitative chemical tests to the use of the most sophisticated and expensive computer-controlled instruments. It is essential that students of the subject gain experience of the more widely-used techniques found in industrial, research and other analytical laboratories. A theme that runs through the practice of analytical chemistry is the need for a careful and disciplined approach, with close attention to detail at all stages of an analytical procedure. Only by adopting this approach can reliable results be obtained, results upon which the economic health or reputation of a company, the clinical health of an individual, or the advancement of research may depend.

The experiments included in this book have been selected to illustrate the use of well-established instrumental techniques that are now the mainstay of modern analytical chemistry, as well as the so-called 'classical' techniques of titrimetry (volumetric analysis) and gravimetry (analysis by weight). Many analytical laboratories cannot justify the purchase of expensive instrumentation and therefore have to rely on the use of relatively simple instruments and on classical and other largely non-instrumental methods of analysis. The selected experiments therefore include a number that require the minimum of specialized equipment and instrumentation (generally costing less than £1000); these are prefixed with the letter 'A'. A further group of experiments, prefixed with the letter 'B', require moderately-priced instruments, i.e. those costing up to about £12,000. Experiments requiring more complex and costly instrumentation have been omitted deliberately from the book in an attempt to ensure the widest appeal. This does not, however, preclude the incorporation of such experiments in possible future editions.

All the experiments are written in a common format that includes a set of grade discussion questions. These form an integral part of each experiment and should not be regarded as optional, although it is recognized that less advance students will need to omit some questions. Most experiments will take average students about three hours. If time is a limiting factor, reagents and even sample solutions can be prepared beforehand by laboratory technical staff.

General Laboratory Practice

It cannot be emphasized too strongly that students of analytical chemistry should adopt a more disciplined approach to common laboratory operations such as weighing, the preparation and transfer of solutions and the control of experimental conditions than is normally necessary for preparative chemistry. Analytical data, by its very nature, must be collected under rigorously controlled and often clinical conditions if confidence is to be placed in the results. The effects of dirty glassware, contaminated reagents or a poor practical technique can easily lead to the generation of misleading or erroneous data. There are many guidelines to good laboratory practice in the context of analytical chemistry and it is worth listing the more important ones so as to encourage the development of a sound practical technique. Students of the subject therefore should pay particular attention to the following:

ALWAYS FOLLOW PROCEDURAL INSTRUCTIONS EXPLICITLY

RECORD DATA AND OBSERVATIONS DIRECTLY INTO A NOTEBOOK - NOT ON
SCRAPS OF PAPER

WEIGH CHEMICALS IN GLASS OR PLASTIC CONTAINERS OR ON GLAZED PAPER

ENSURE THAT SAMPLES, STANDARDS AND REAGENTS ARE LABELLED

ALWAYS USE CLEAN GLASSWARE

NEVER HEAT CALIBRATED GLASSWARE

USE ANALAR QUALITY REAGENTS UNLESS OTHERWISE INSTRUCTED

These are reagents with guaranteed maximum limits of impurities.

TAKE CARE TO AVOID ACCIDENTAL CONTAMINATION OF STANDARDS,
SAMPLES AND REAGENTS

BEWARE OF VARIATIONS IN THE QUALITY OF REAGENT CHEMICALS AND SOLVENTS

MAKE DUPLICATE MEASUREMENTS AND RUN DUPLICATE ANALYSES WHEREVER POSSIBLE

CRITICALLY EVALUATE ALL MEASUREMENTS AND REJECT ANY THAT ARE SUSPECT

USE STATISTICAL METHODS TO EVALUATE QUANTITATIVE DATA

Preparation and Handling of Sample and Standard Solutions

Whenever possible, the minimum volumes of solutions should be prepared, as many solvents and reagent chemicals are expensive and/or toxic. Always dissolve substances in the minimum of solvent in a beaker covered with a watch glass, warming or boiling if necessary to aid dissolution. After cooling to room temperature, it is advisable to place a small funnel into the neck of the volumetric flask before transferring the solution quantitatively from the beaker. This minimizes the chances of spillage. The watch glass and the sides of the beaker should be rinsed several times with small portions of solvent or water, adding the rinsings to the contents of the flask before diluting to volume.

N.B. NEVER HEAT SOLUTIONS IN VOLUMETRIC FLASKS

They are accurately calibrated and may be adversely affected by such treatment.

Dilute standards should always be prepared by dilution of more concentrated stock solutions. This is a more accurate procedure than weighing very small quantities. Furthermore, very dilute solutions should be prepared fresh and not used after they are more than a day or so old. This is because of the risk of significant losses by adsorption on or interaction with the walls of the container. Plastic storage bottles are more suitable than glass in this respect.

All volumetric glassware should be kept scrupulously clean; significant variations in the volumes delivered and contained may occur if grease or other residues are not removed by thorough cleaning. Treatment with a nitric/chromic acid mixture followed by several rinsings with distilled or deionized water is recommended for this purpose.

INTRODUCTION

Calibration Procedures Used in Quantitative Analysis

Very few quantitative analytical methods are absolute in the sense that the
measured parameter leads directly to the mass, volume or concentration of
the analyte* in question. In most cases, the system needs to be calibrated
by measuring the response of a standard or the responses of a series of
standards of known composition with respect to the analyte. These are then
compared with the response from the analyte in a sample to establish the
amount present. The comparison may be made by reference to a previously
prepared calibration curve or by computation, the latter having become
more common with the wide availability of microprocessor-controlled and
computerized instrumentation.

The following four methods are widely used in quantitative analysis, each
one having merits that make it suitable for particular types of sample and/or
analytical requirements.

1. CALIBRATION CURVES are graphs of the mass, volume or concentration of an
analyte in standards of known composition plotted as a function of instrumental
response, e.g. absorbance or emission intensity of electromagnetic radiation,
electrochemical cell potential or current, chromatographic peak area etc. In
cases where the sample matrix or reagents contribute significantly to the
response, this contribution, known as a 'blank', is substracted from the
total response in each case to give the response of the analyte alone. Data is
usually plotted directly, but in some cases logarithmic functions or ratios are
used. A linear relation between the variables plotted is normally sought, but
curved calibration graphs may be acceptable if the cause of the curvature is
understood and is reproducible. Some modern instruments employ automatic
curvature-correction routines to extend the linear range of calibration data.

* Constituent of the sample (element or compound) that is to be determined
quantitatively or identified qualitatively.

Where a linear relation is established but the scatter of points is
appreciable, linear regression analysis (p.16) should be used to obtain a line
of best fit. Calibration graphs are easy to use as the amount or
concentration of the analyte in a sample is simply read from the curve or
computed using a factor where the curve is linear.

2. STANDARD ADDITION is a technique used to prepare a calibration graph in
cases where the composition of the sample matrix is variable or unknown so
that a reagent/sample matrix 'blank' response cannot reliably be subtracted
from each standard to arrive at the analyte response alone (see 1). In
such cases, a series of samples are 'spiked' with different known amounts of
the analyte. The responses of the 'unspiked' and 'spiked' samples are then
measured and a calibration curve plotted as shown below.

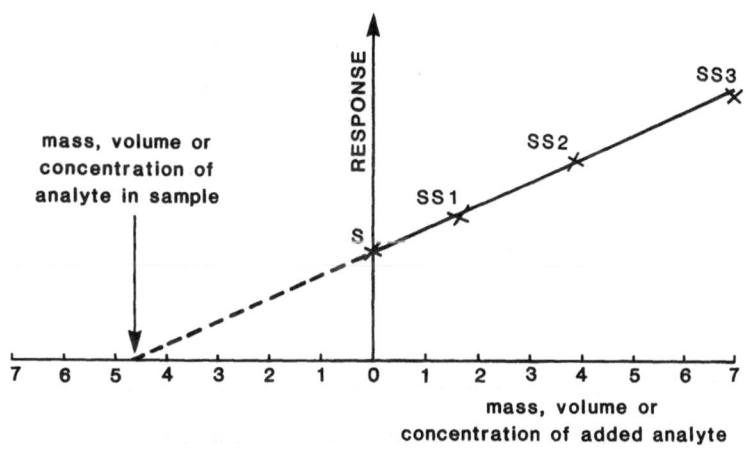

Figure 1. Calibration curve for standard addition

The responses of at least three samples spiked with different amounts of the
analyte (one only if time is limited) together with that of the unspiked
sample are plotted as ordinates against the corresponding masses, volumes or
concentrations as abscissae. The curve, which will NOT pass through the
origin, is extrapolated backwards to the point where it meets a 'mirror image'
of the abscissa axis. The abscissa scale reading at this point gives the

mass, volume or concentration of the analyte in the sample provided that no
significant dilutions of the samples occurred on spiking or that the unspiked
and spiked samples were diluted by the same amount. If this is not the case,
appropriate corrections must be applied to the responses.

3.   INTERNAL STANDARDIZATION involves the addition of a constant mass,
volume or concentration of a selected standard (not the analyte itself) to
all samples and to one or more analyte standards. For each sample and
analyte standard, the responses of the analyte and the added internal standard
are measured and the RATIOS of these responses calculated. If several
analyte standards are treated in this way, a calibration curve can be plotted
of RESPONSE RATIOS against the mass, volume or concentration of the analyte.
Note that the response ratios are or should be independent of the absolute
values of the responses themselves, i.e. of the sample size. The
compositions of samples can then be read from the calibration curve using
their calculated response ratios. If only one or two analyte standards are
prepared, the sample composition is established arithmetically by simple
proportion, i.e.

$$\frac{\text{mass, volume or concentration of analyte in sample}}{\text{mass, volume or concentration of analyte in standard}} = \frac{\text{response ratio for sample}}{\text{response ratio for analyte standard}}$$

The particular value of internal standardization is that response ratios
are ideally independent of sample size and are usually much less sensitive to
variations in experimental conditions than the response of the analyte
itself.

4.   INTERNAL NORMALIZATION is used when only the proportions of two or more
analytes in a sample are to be determined. The proportions can be
normalized to any number, but 100 per cent, or 1 are the most usual. The
composition of the sample is calculated by expressing the response for each
component, corrected for differences in sensitivity if necessary, as a
percentage or fraction of the summed responses for all the components. Thus
the internally normalized composition of a three-component mixture

separated by gas or liquid chromatography would be established by measuring the peak area for each and expressing it as a percentage or fraction of the summed areas, e.g.

$$\% \text{ of component A} = \frac{\text{peak area for component A}}{\text{sum of areas for A,B and C}} \times 100 \text{ etc.}$$

Units of Concentration

The concentration of standards and reagent solutions used in analytical procedures and quantitative results are expressed in a variety of units. It is essential to be familiar with the units in common usage and with their interrelations. In particular, the relation between parts per million (ppm) and concentrations expressed in micrograms (μg), milligrams (mg), grams (g) and per cent, should be thoroughly understood.
Molarity is expressed in moles per cubic decimetre, mol dm^{-3}, although in experimental instructions it is convenient to retain the older abbreviation 'M' for common bench reagents such as acids and alkalis as this is how they are identified in most laboratories.

Concentration units are summarized in Table 1.

Table 1 Alternative methods of expressing concentration*

UNITS	NAME AND SYMBOL
moles of solute per dm^3	$mol\ dm^{-3}$, M
equivalents of solute per dm^3	normal, N
milli-equivalents of solute per dm^3	$meq\ dm^{-3}$
grams of solute per dm^3	$g\ dm^{-3}$
parts per million	ppm (γ)
milligrams of solute per dm^3	$mg\ dm^{-3}$
parts per billion	ppb
nanograms of solute per dm^3	$ng\ dm^{-3}$
parts per hundred	% (w/w, w/v, v/v)
millimoles of solute per 100 cm^3	mM %
grams of solute per 100 cm^3	g %
milligrams of solute per 100 cm^3	mg %
micrograms of solute per 100 cm^3	μg %
nanograms of solute per 100 cm^3	ng %
micrograms of solute per cm^3	$\mu g\ cm^{-3}$ } $\equiv$ ppm
micrograms per gram	$\mu g\ g^{-1}$
nanograms of solute per cm^3	$ng\ cm^{-3}$
nanograms per gram	$ng\ g^{-1}$
picograms of solute per cm^3	$pg\ cm^{-3}$ } $\equiv$ ppb
picograms per gram	$pg\ g^{-1}$

*The table includes most of the methods of expressing concent-
ration that are in current use, although some are not consistent
with S.I.

Some Useful Relations and Definitions

The Beer-Lambert Law

This law relates the absorbance of electromagnetic radiation by solute
species in solution to their concentration and to the path length of the
sample cell. It is used to prepare calibration curves (p. 4) for quantitative
analysis by ultraviolet, visible and infrared spectrometry and is expressed in
the form

$$A = \epsilon b c$$

where A is the absorbance of the solute (A= $-\log_{10}T$, where T is the

transmittance)

ϵ is the molar absorptivity of the solute (a constant defined as the

absorbance of a 1 mol dm^{-3} solution in a 1-cm cell, or in SI units

the absorbance of a 1 mol m^{-3} solution in a 1-m cell)

b is the path length of the cell in cm or m in SI units

c is the concentration of the solute in mol dm-3, or mol m^{-3} in SI

units

In some instances, e.g. when the relative molecular mass (RMM) of the solute
is unknown, ϵ is replaced by an alternative absorptivity constant, E_y^x, where
x defines the path length of the cell in cm and y is the concentration of
solute, measured in per cent, weight per unit volume or some other convenient
concentration units.

For systems that obey the Beer-Lambert law, a plot of A against c will be
linear and should pass through the origin, the slope of the curve being given
by the product $\epsilon.b$ or $E_y^x.b$.
(Details of the derivation of the law and limitations in its use are given in
most standard texts on analytical chemistry).

The Nernst Equation

This equation relates the concentration (strictly the activity) of an analyte

INTRODUCTION

ion in solution to the potential of an electrochemical cell comprising an indicator electrode sensitive to the analyte ion, a reference electrode and the analyte solution. It is used in quantitative measurements by direct potentiometry and as the basis of potentiometric titrations. It may be written in the form

$$E_{cell} = k - \frac{0.059V}{n} \log_{10} a_x$$

where E_{cell} is the measured cell potential in V or mV

    k is a constant that includes the reference electrode potential

    n is the number of electrons involved in the indicator electrode reaction (for ion selective electrodes this is taken to be the formal valency of the analyte ion)

    a_x is the concentration (strictly activity) of the analyte ion

A plot of E_{cell} (read from a millivolt meter) against $\log_{10} a_x$ will be linear with a slope of $- \frac{0.059V}{n}$ or $- \frac{59 \text{ mV}}{n}$, i.e. there is a change of $\frac{59}{n}$ mV per 10 - fold change in concentration of the analyte ion. In practice, the slope may vary by several millivolts from that predicted by the Nernst equation and curvature may be observed at very low concentrations. Nernst equation graphs are used as calibration curves (p. 4)in quantitative analysis by potentiometry. Plotting the curve is facilitated by the use of log-linear (semi-log) graph paper where E_{cell} is plotted on the linear scale and a_x on the log scale.

Chromatographic Analysis

Substances separated by chromatographic techniques are characterized by their rate of migration through the stationary phase. For column separations, this is related to the RETENTION time, t_R, defined as the elapsed time between injection of the sample on to the column and its emergence from the other end. It is normally measured from the chromatogram (chart record) as a length in cm or mm that is directly proportional to t_R (see Figure 2 below). For thin-layer separations, substances are characterized by their R_f value, defined as

$$\frac{\text{distance travelled by substance}}{\text{distance travelled by solvent front}}$$

the values ranging from zero to one.

The quality of chromatographic performance is assessed in terms of
EFFICIENCY and RESOLUTION. EFFICIENCY is a measure of the degree of peak
broadening that occurs during a separation and is expressed in terms of the
PEAK WIDTH and the RETENTION TIME of a chromatographed substance.
Two alternative formulae are used for calculating efficiency from a
chromatogram, viz.

$$(1) \quad N = 16(t_R/W_b)^2 \qquad \text{and} \qquad (2) \quad N = 5.54 \ (t_R/W_{h/2})^2$$

where N is the efficiency or NUMBER OF THEORETICAL PLATES generated
    by the substance

t_R is the RETENTION TIME of the substance

W_b is the BASELINE PEAK WIDTH

$W_{h/2}$ is the PEAK WIDTH AT ONE-HALF OF THE PEAK HEIGHT

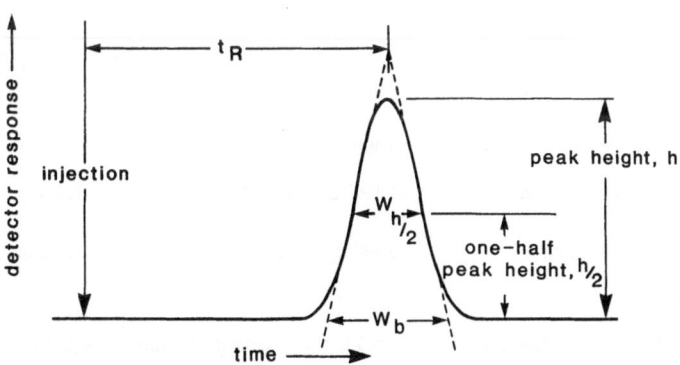

Figure 2   The measurement of chromatographic
retention time and efficiency

Formula (1) is used widely but suffers from the difficulty of accurately
locating the inflection points on each side of the peak and in drawing the
tangents to these points which define W_b , the base width. Formula (2) is more

satisfactory in that only the peak height and $W_{h/2}$, the width at half the peak height, need to be measured. However, the accuracy of measuring $W_{h/2}$ diminishes as the peak narrows. This formula is less affected by peak tailing than formula (1).

> N.B.   All comparisons of chromatographic efficiency
> should be made using the same formula as the two
> generally give different plate numbers for the
> same peak.

Efficiency may also be quoted in terms of the HEIGHT EQUIVALENT OF A THEORETICAL PLATE, H, which is related to the plate number, N, by the equation

$$H = L/N$$

> where L is the length of the chromatographic column, normally
> quoted in mm or cm.
> H has the advantage of being independent of column length.

Typical values for N and H for gas and liquid chromatography are given below.

	Total plate number,N	Plate height,H
Gas chromatography		
packed column	1000 to 5000	0.5 mm
capillary column	50000 to 500000	0.4 mm
Liquid chromatography	1000 to 8000	0.03 mm

RESOLUTION   is a measure of how well two substances have been separated and is defined by the expression

$$R_S = \frac{2 \ \Delta t_R}{(W_1 + W_2)}$$

> where R_S is the measured RESOLUTION (see Figure 3)
> Δt_R is the peak-to-peak separation of the two substances and
> W_1 and W_2 are the respective BASE WIDTHS of the two peaks.

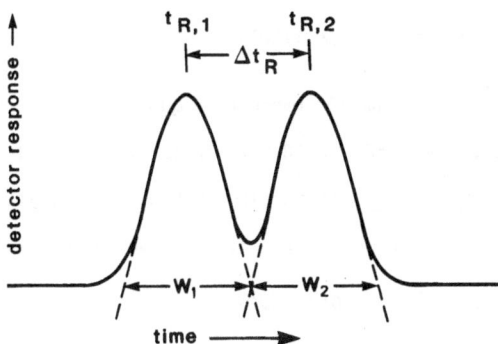

Figure 3   The measurement of chromatographic
            resolution

An R_S value of 1.5 or greater is indicative of complete or 'baseline'
separation when the peaks are of approximately the same size. Quantitative
analysis by peak area measurement is generally not possible if the R_S value
is below 0.8.

Significant Figures

In reporting numerical results, it is common practice to round them off to a
selected number of SIGNIFICANT FIGURES. The convention is to include
all digits that are known with certainty followed by the first that is
uncertain. Zeros are significant only if they form part of the number and
not when they merely indicate magnitude, i.e. 9.032 and 0.009032 both
contain four significant figures (assuming that the 2 is the first uncertain
digit). The assessment of which is the first uncertain digit in a result
should be made by careful consideration of the errors associated with each
number used in its derivation. Sometimes this can be done only with
extensive knowledge of the analytical procedure and the nature of the
measurements involved. In general, the uncertainty in a derived result will
be determined by the largest absolute or relative uncertainty associated with

the numbers used to calculate the result, e.g.

1. When adding or subtracting 155.5± 0.1 and 0.085 ± 0.001, the result is 155.585 or 155.415. The largest absolute uncertainty appears in the fourth digit, whence the result should be rounded off to 155.6 or 155.4 respectively.

2. When calculating the product or quotient of 155.5 ± 0.1 and 0.085 ± 0.001, the result is 13.2175 or 1829.411764. The largest relative uncertainty appears in the third digit, whence the result should be rounded off to 13.2 or 1830 respectively.

Statistical Assessment of Quantitative Data

Whenever quantitative determinations are made, it is important to establish the quality or reliability of the results. This is particularly true in industrial or other analytical laboratories where those results relate to the cost of a raw material, the quality of a product or the condition of a patient. In devising and testing new methods of analysis, it is usual to compare the results with those obtained from other methods, if they exist, and often with results obtained by different analysts and in different laboratories, i.e. collaborative testing.

The quality of quantitative data can be described in terms of its PRECISION and ACCURACY. Both are assessed by statistical means using formulae derived for the purpose of treating small sets of data. PRECISION is defined as the reproducibility or variability of a set of results or measurements that have all been made in an IDENTICAL FASHION. It is a measure of the effect of INDETERMINATE (RANDOM) errors on the measurement or result. ACCURACY is defined as the closeness of a result or measurement (or the mean of a set of replicates) to a TRUE or ACCEPTED VALUE. A significant departure from the true or accepted value is indicative of a BIAS in the procedure. PRECISION is normally assessed by estimating the STANDARD DEVIATION of a set of replicate results or measurements. Often, it is expressed as the RELATIVE PRECISION (RELATIVE STANDARD DEVIATION OR COEFFICIENT OF VARIATION), which is a percentage figure that is independent of the units in which the results are expressed (Formulae are given in the table on pp. 17 and 18.)

A comparison of the precisions of two sets of data can be made using
the F-TEST, where F_{expt}, the ratio of the experimental VARIANCES (squares
of the estimated standard deviations) for the two sets, is compared with a
tabular value, F_{tab}. If the experimental ratio exceeds the tabular value,
at a selected level of confidence or probability (usually 95%), then the
precisions of the two sets of data are considered to be significantly
different.

ACCURACY is assessed by the calculation of a CONFIDENCE INTERVAL about
an experimental value or the mean of a set of values. The interval defines
a range within which the true or accepted value is expected to lie with a
selected degree of confidence (probability), usually 90% or 95%. It is
derived from a formula incorporating the estimated standard deviation, the
number of results used to estimate it and an appropriate tabular value of
the t-FACTOR.

A comparison of the experimental means derived from two sets of data, e.g.
results from two laboratories, analysts or methods, or of an experimental
mean with a true or accepted value, can be made using the t-TEST. Where
two sets of experimental data are involved, the comparison is valid only if
an F-test has shown that the precisions of the two sets are not
significantly different. An experimental t-value is first calculated from
the means, or from a mean and the true or accepted value, the standard
deviation (a pooled value is used where two sets of data are involved), and
the number(s) of results in the set(s). This is then compared with a
tabular t-value at a selected confidence (probability) level, usually 90%
or 95%. If the experimental value exceeds the tabular value, the means are
considered to be significantly different, or the experimental mean is
considered to differ significantly from the true or accepted value (Formulae
are given in the table on pp. 17 and 18.)

Other statistical methods are used for assessing the quality or reliability
of calibration data and the effect of experimental variables on measured
responses. Two of particular value are LINEAR REGRESSION ANALYSIS and the
calculation of a CORRELATION COEFFICIENT.

INTRODUCTION

Calibration curves are used widely in quantitative analysis (see p. 4) and are frequently linear. However, indeterminate errors associated with the measurement process often result in a series of plotted points that do not exactly represent a linear relation, i.e. a straight line may not pass through all points. It may be possible to judge the position of the 'best' straight line by eye, but if the scatter of points is appreciable LINEAR REGRESSION ANALYSIS will prove more reliable. This involves computing a slope and an intercept that define a REGRESSION LINE (line of best fit), the line representing the average relation between the two plotted variables. As well as providing a working calibration curve, the degree of scatter of the plotted points around the REGRESSION LINE indicates the precision of the calibration data and is normally quoted as a standard deviation about the line. The regression line can be defined using the "method of least squares", the appropriate formulae being given in the table on p.18 and a simple computer program will enable the computations to be made rapidly and reliably.

It is usually necessary during the development of a new analytical method to establish whether there is a linear relation between, for example, the mass, volume or concentration of the analyte and a measured instrumental response, or if the response is affected by another variable in the system, e.g. temperature, excess reagent etc. The preparation of a calibration curve (see p. 4) is the most straightforward approach, but the calculation of a CORRELATION COEFFICIENT, r, based on a series of paired values, e.g. concentration and response, enables the degree of linearity (correlation) to be assessed quantitatively. For a perfectly linear relation, the value of r will be ± 1. Values that exceed ± 0.7 indicate acceptable linearity, whilst values below ± 0.5 are evidence of little correlation between the two variables. The equation defining the correlation coefficient is given in the table on p. 18.

The statistical formulae used most frequently in assessing the quality of quantitative data are given in the following table. Tables of F and t values are given in Appendices 4 and 5. More detailed treatments of the use of statistical methods in analytical chemistry are given in various textbooks (see references 1 to 3).

Table 2   Statistical Formulae

FORMULA	USE
Estimated Standard Deviation	
$$s = \left[\frac{\sum\limits_{i=1}^{i=N} (x_i - \bar{x})^2}{N - 1}\right]^{\frac{1}{2}}$$	To estimate the standard deviation of a set of replicate measurements or results

s = estimated standard deviation

N = number of replicate measurements or results

x_i = individual measurement or result

$\bar{x}$ = arithmetic mean of the set

Pooled Standard Deviation	To estimate the pooled standard deviation based on data from more than one set of replicate measurements or results
$$s_{pooled} = \left[\frac{\sum\limits_{i=1}^{N_1}(x_i-\bar{x}_1)^2 + \sum\limits_{i=1}^{N_2}(x_i-\bar{x}_2)^2 + \ldots \sum\limits_{i=1}^{N_K}(x_i-\bar{x}_K)^2}{M - K}\right]^{\frac{1}{2}}$$	

s_{pooled} = estimated pooled standard deviation

$N_1, N_2 \ldots N_K$ = number of replicate measurements or results in each of K sets

M = total number of replicate measurements or results

x_i = individual measurement or result

$\bar{x}_1, \bar{x}_2 \ldots \bar{x}_K$ = arithmetic means of each of K sets

Relative Standard Deviation	
$$s_r = \frac{s}{\bar{x}} \cdot 100$$	To calculate the relative precision for comparative purposes

s_r = relative standard deviation, relative precision or coefficient of variation

s = estimated standard deviation

$\bar{x}$ = arithmetic mean of the set

F-test	To compare the precision of two sets of data using F-test tables
$$F_{expt} = \frac{s_1^2}{s_2^2}$$	

s_1 = estimated standard deviation for first set

s_2 = estimated standard deviation for second set

N.B. $s_1 > s_2$

F_{tab} = tabular F-value with appropriate number of degrees of freedom for numerator and denominator

Confidence Interval	
$$C.I. = \bar{x} \pm ts/N^{\frac{1}{2}}$$	To establish the confidence interval around an experimental mean, i.e. an estimate of accuracy

C.I. = confidence interval at selected level of confidence

$\bar{x}$ = arithmetic mean of the set of results

s = estimated standard deviation

N = number of replicate results

t = tabular t-value at selected level of confidence
and N-1 degrees of freedom

t-test

$$t_{expt} = \left[(\bar{x}-\mu)/s \right] N^{\frac{1}{2}}$$

To compare the mean of a set of results with a true or accepted value, i.e. to establish the reliability of the experimental mean

$\bar{x}$ = arithmetic mean of the set of results

μ = true or accepted mean

s = estimated standard deviation

N = number of replicate results

$$t_{expt} = \left[(\bar{x}-\bar{y})/s_{pooled} \right] \left[MN/(M+N) \right]^{\frac{1}{2}}$$

To compare the means of two sets of results, i.e. to establish the level of agreement between two experimental means

$\bar{x}$ = arithmetic mean of first set of results

$\bar{y}$ = arithmetic mean of second set of results

s_{pooled} = estimated pooled standard deviation

M = number of replicate results in first set

N = number of replicate results in second set

Linear Regression Equations

$$y = bx + a$$

Expresses a linear relation between two variables x and y

$$b = \frac{\Sigma y_i \Sigma x_i - n \Sigma x_i y_i}{(\Sigma x_i)^2 - n\Sigma x_i^2}$$

To calculate the slope of the regression line for n pairs of values of x and y

$$a = \frac{\Sigma x_i \Sigma x_i y_i - \Sigma y_i \Sigma x_i^2}{(\Sigma x_i)^2 - n\Sigma x_i^2}$$

To calculate the intercept of the regression line for n pairs of values of x and y

$$s = \left[\frac{\Sigma (y_i^{calc.} - y_i^{expt})^2}{n - 2} \right]^{\frac{1}{2}}$$

To calculate the standard deviation about the regression line for n pairs of values of x and y

Correlation Coefficient

$$r = \frac{\Sigma (x_i-\bar{x}) (y_i-\bar{y})}{\left[\Sigma(x-\bar{x})^2 . \Sigma(y-\bar{y})^2 \right]^{\frac{1}{2}}}$$

To check the degree of correlation (linearity) between two variables x and y

Appendix 1   International atomic weights, 1973

Element	Symbol	Atomic No.	Atomic Weight	Element	Symbol	Atomic No.	Atomic Weight
Actinium	Ac	89	(227)	Mercury	Hg	80	200.59
Aluminium	Al	13	26.9815	Molybdenum	Mo	42	95.94
Americium	Am	95	(243)	Neodymium	Nd	60	144.24
Antimony	Sb	51	121.75	Neon	Ne	10	20.179
Argon	Ar	18	39.948	Neptunium	Np	93	237.0482
Arsenic	As	33	74.916	Nickel	Ni	28	58.70
Astatine	At	85	(210)	Niobium	Nb	41	92.9064
Barium	Ba	56	137.34	Nitrogen	N	7	14.0067
Berkelium	Bk	97	(247)	Nobelium	No	102	(255)
Beryllium	Be	4	9.0122	Osmium	Os	76	190.2
Bismuth	Bi	83	208.9804	Oxygen	O	8	15.9994
Boron	B	5	10.81	Palladium	Pd	46	106.4
Bromine	Br	35	79.904	Phosphorus	P	15	30.9738
Cadmium	Cd	48	112.40	Platinum	Pt	78	195.09
Calcium	Ca	20	40.08	Plutonium	Pu	94	(244)
Californium	Cf	98	(251)	Polonium	Po	84	(209)
Carbon	C	6	12.011	Potassium	K	19	39.098
Cerium	Ce	58	140.12	Praseodymium	Pr	59	140.9077
Cesium	Cs	55	132.9054	Promethium	Pm	61	(145)
Chlorine	Cl	17	35.453	Protactinium	Pa	91	231.0359
Chromium	Cr	24	51.996	Radium	Ra	88	226.0254
Cobalt	Co	27	58.9332	Radon	Rn	86	(222)
Copper	Cu	29	63.546	Rhenium	Re	75	186.207
Curium	Cm	96	(247)	Rhodium	Rh	45	102.9055
Dysprosium	Dy	66	162.50	Rubidium	Rb	37	85.4678
Einsteinium	Es	99	(254)	Ruthenium	Ru	44	101.07
Erbium	Er	68	167.26	Samarium	Sm	62	150.4
Europium	Eu	63	151.96	Scandium	Sc	21	44.9559
Fermium	Fm	100	(257)	Selenium	Se	34	78.96
Fluorine	F	9	18.9984	Silicon	Si	14	28.086
Francium	Fr	87	(223)	Silver	Ag	47	107.868
Gadolinium	Gd	64	157.25	Sodium	Na	11	22.9898
Gallium	Ga	31	69.72	Strontium	Sr	38	87.62
Germanium	Ge	32	72.59	Sulphur	S	16	32.06
Gold	Au	79	196.9665	Tantalum	Ta	73	180.9479
Hafnium	Hf	72	178.49	Technetium	Tc	43	(97)
Helium	He	2	4.0026	Tellurium	Te	52	127.60
Holmium	Ho	67	164.9304	Terbium	Tb	65	158.9254
Hydrogen	H	1	1.0079	Thallium	Tl	81	204.37
Indium	In	49	114.82	Thorium	Th	90	232.0381
Iodine	I	53	126.9045	Thulium	Tm	69	168.9342
Iridium	Ir	77	192.22	Tin	Sn	50	118.69
Iron	Fe	26	55.847	Titanium	Ti	22	47.90
Krypton	Kr	36	83.80	Tungsten	W	74	183.85
Lanthanum	La	57	138.9055	Uranium	U	92	238.029
Lawrencium	Lr	103	(260)	Vanadium	V	23	50.9414
Lead	Pb	82	207.2	Xenon	Xe	54	131.30
Lithium	Li	3	6.941	Ytterbium	Yb	70	173.04
Lutetium	Lu	71	174.97	Yttrium	Y	39	88.9059
Magnesium	Mg	12	24.305	Zinc	Zn	30	65.38
Manganese	Mn	25	54.9380	Zirconium	Zr	40	91.22
Mendelevium	Md	101	(258)				

Notes:

1. This table is scaled to the relative atomic mass $A_r(^{12}C) = 12$.

2. Values in parentheses refer to the isotope of longest known half-life for radioactive elements.

3. Information provided here is based upon the Report of the Commission on Atomic Weights, Pure and Applied Chemistry, (1974), 37, 589.

APPENDIX 2

Appendix 2   Dissociation constants of some acids in water at 25°C

Dissociation constants are expressed as $pK_a (= -\log K_a)$.

Acid		pK_a	Acid		pK_a
Aliphatic acids					
Acetic		4.76	Succinic	K_1	4.21
Propanoic		4.87		K_2	5.64
Butanoic		4.82	Glutaric	K_1	4.34
3-Methyl propanoic		4.85		K_2	5.27
Pentanoic		4.84	Adipic	K_1	4.43
Chloroacetic		2.86		K_2	5.28
Bromoacetic		2.90	Methylmalonic	K_1	3.07
Diethylacetic		4.73		K_2	5.87
Lactic		3.86	Ethylmalonic	K_1	2.96
Pyruvic		2.49		K_2	5.90
Acrylic		4.26	Dimethylmalonic	K_1	3.15
Vinylacetic		4.34		K_2	6.20
Furoic		3.17	Diethylmalonic	K_1	2.15
Oxalic	K_1	1.27		K_2	7.47
	K_2	4.27	Fumaric	K_1	3.02
Malonic	K_1	2.85		K_2	4.38
	K_2	5.70	Maleic	K_1	1.92
				K_2	6.23
			Tartaric	K_1	3.03
				K_2	4.37
			Citric	K_1	3.13
				K_2	4.76
				K_3	6.40
Aromatic acids					
Benzoic		4.21	2-Benzoylbenzoic		3.54
Phenylacetic		4.31	Phthalic	K_1	2.95
Sulphanilic		3.23		K_2	5.41
Phenoxyacetic		3.17	cis-Cinnamic		3.88
Mandelic		3.41	trans-Cinnamic		4.44
1-Naphthoic		3.70	Phenol		10.00
2-Naphthoic		4.16			
1-Naphthylacetic		4.24			
2-Naphthylacetic		4.26			

Appendix 2   continued

Some typical buffer solutions

SOLUTIONS	pH RANGE
phthalic acid and potassium hydrogen phthalate	2.2-4.2
citric acid and sodium citrate	2.5-7.0
acetic acid and sodium acetate	3.8-5.8
sodium dihydrogen phosphate and disodium hydrogen phosphate	6.2-8.2
ammonia and ammonium chloride	8.2-10.2
borax and sodium hydroxide	9.2-11.2

A range of visual indicators for acid-base titrations

INDICATOR	pK_{In}	LOW pH COLOUR	HIGH pH COLOUR	EXPERIMENTAL COLOUR CHANGE RANGE/pH
cresol red	ca.1	red	yellow	0.2-1.8
thymol blue	1.7	red	yellow	1.2-2.8
bromo-phenol blue	4.0	yellow	blue	2.8-4.6
methyl orange	3.7	red	yellow	3.1-4.4
methyl red	5.1	red	yellow	4.2-6.3
bromo-thymol blue	7.0	yellow	blue	6.0-7.6
phenol red	7.9	yellow	red	6.8-8.4
phenolphthalein	9.6	colourless	red	8.3-10.0
alizarin yellow R	ca.11	yellow	orange	10.1-12.0
nitramine	ca.12	colourless	orange	10.8-13.0

Appendix 3   Some selected metal-EDTA formation constants

CATION	K_{MY}	$\log_{10}K_{MY}$	CATION	K_{MY}	$\log_{10}K_{MY}$
Ag^+	2×10^7	7.3	Cu^{2+}	6.3×10^{18}	18.80
Mg^{2+}	4.9×10^8	8.69	Zn^{2+}	3.2×10^{16}	16.50
Ca^{2+}	5.0×10^{10}	10.70	Cd^{2+}	2.9×10^{16}	16.46
Sr^{2+}	4.3×10^8	8.63	Hg^{2+}	6.3×10^{21}	21.80
Ba^{2+}	5.8×10^7	7.76	Pb^{2+}	1.1×10^{18}	18.04
Mn^{2+}	6.2×10^{13}	13.79	Al^{3+}	1.3×10^{16}	16.13
Fe^{2+}	2.1×10^{14}	14.33	Fe^{3+}	1×10^{25}	25.1
Co^{2+}	2.0×10^{16}	16.31	V^{3+}	8×10^{25}	25.9
Ni^{2+}	4.2×10^{18}	18.62	Th^{4+}	2×10^{23}	23.2

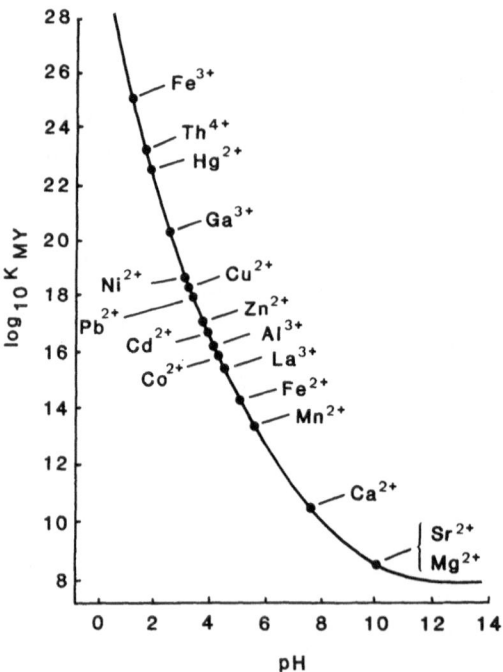

Minimum pH needed for the satisfactory
titration of various cations with EDTA as
a function of the logarithm of the complex
formation constant

Appendix 4   Table of F-Distribution

CONFIDENCE (PROBABILITY) LEVEL, %	DEGREES OF FREEDOM DENOMINATOR	NUMERATOR											
		1	2	3	4	5	6	7	8	9	10	15	∞
90	1	39.9	49.5	53.6	55.8	57.2	58.2	58.9	59.4	59.9	60.2	61.2	63.3
95		161.4	199.5	215.7	224.6	230.2	234.0	236.8	238.9	240.5	241.9	246.0	254.3
99		4,052	4,999	5,403	5,625	5,764	5,859	5,928	5,981	6,023	6,056	6,157	6,366
90	2	8.53	9.00	9.16	9.24	9.29	9.33	9.35	9.37	9.38	9.39	9.42	9.49
95		18.5	19.0	19.2	19.2	19.3	19.3	19.4	19.4	19.4	19.4	19.4	19.5
99		98.5	99.0	99.2	99.3	99.3	99.3	99.4	99.4	99.4	99.4	99.4	99.5
90	3	5.54	5.46	5.39	5.34	5.31	5.28	5.27	5.25	5.24	5.23	5.20	5.13
95		10.1	9.55	9.28	9.12	9.01	8.94	8.89	8.85	8.81	8.79	8.70	8.53
99		34.1	30.8	29.5	28.7	28.2	27.9	27.7	27.5	27.3	27.2	26.9	26.1
90	4	4.54	4.32	4.19	4.11	4.05	4.01	3.98	3.95	3.94	3.92	3.87	3.76
95		7.71	6.94	6.59	6.39	6.26	6.16	6.09	6.04	6.00	5.96	5.86	5.63
99		21.2	18.0	16.7	16.0	15.5	15.2	15.0	14.8	14.7	14.5	14.2	13.5
90	5	4.06	3.78	3.62	3.52	3.45	3.40	3.37	3.34	3.32	3.30	3.24	3.10
95		6.61	5.79	5.41	5.19	5.05	4.95	4.88	4.82	4.77	4.74	4.62	4.36
99		16.3	13.3	12.1	11.4	11.0	10.7	10.5	10.3	10.2	10.1	9.72	9.02
90	6	3.78	3.46	3.29	3.18	3.11	3.05	3.01	2.98	2.96	2.94	2.87	2.72
95		5.99	5.14	4.76	4.53	4.39	4.28	4.21	4.15	4.10	4.06	3.94	3.67
99		13.7	10.9	9.78	9.15	8.75	8.47	8.26	8.10	7.98	7.87	7.56	6.88
90	7	3.59	3.26	3.07	2.96	2.88	2.83	2.78	2.75	2.72	2.70	2.63	2.47
95		5.59	4.74	4.35	4.12	3.97	3.87	3.79	3.73	3.68	3.64	3.51	3.23
99		12.2	9.55	8.45	7.85	7.46	7.19	6.99	6.84	6.72	6.62	6.31	5.65
90	8	3.46	3.11	2.92	2.81	2.73	2.67	2.62	2.59	2.56	2.54	2.46	2.29
95		5.32	4.46	4.07	3.84	3.69	3.58	3.50	3.44	3.39	3.35	3.22	2.93
99		11.3	8.65	7.59	7.01	6.63	6.37	6.18	6.03	5.91	5.81	5.52	4.86

Appendix 4   Table of F-Distribution (continued)

CONFIDENCE (PROBABILITY) LEVEL, %	DEGREES OF FREEDOM DENOMINATOR	NUMERATOR											
		1	2	3	4	5	6	7	8	9	10	15	∞
90	9	3.36	3.01	2.81	2.69	2.61	2.55	2.51	2.47	2.44	2.42	2.34	2.16
95		5.12	4.26	3.86	3.63	3.48	3.37	3.29	3.23	3.18	3.14	3.01	2.71
99		10.6	8.02	6.99	6.42	6.06	5.80	5.61	5.47	5.35	5.26	4.96	4.31
90	10	3.29	2.92	2.73	2.61	2.52	2.46	2.41	2.38	2.35	2.32	2.24	2.06
95		4.96	4.10	3.71	3.48	3.33	3.22	3.14	3.07	3.02	2.98	2.85	2.54
99		10.0	7.56	6.55	5.99	5.64	5.39	5.20	5.06	4.94	4.85	4.56	3.91
90	15	3.07	2.70	2.49	2.36	2.27	2.21	2.16	2.12	2.09	2.06	1.97	1.76
95		4.54	3.68	3.29	3.06	2.90	2.79	2.71	2.64	2.59	2.54	2.40	2.07
99		8.68	6.36	5.42	4.89	4.56	4.32	4.14	4.00	3.89	3.80	3.52	2.87
90	∞	2.71	2.30	2.08	1.94	1.85	1.77	1.72	1.67	1.63	1.60	1.49	1.00
95		3.84	3.00	2.60	2.37	2.21	2.10	2.01	1.94	1.88	1.83	1.67	1.00
99		6.63	4.61	3.78	3.32	3.02	2.80	2.64	2.51	2.41	2.32	2.04	1.00

Appendix 5   Table of t-Distribution

CONFIDENCE (PROBABILITY) LEVEL, %

DEGREES OF FREEDOM	80	90	95	98	99	99.5	99.9
1	3.08	6.31	12.71	31.82	63.66	127.32	636.62
2	1.89	2.92	4.30	6.97	9.92	14.09	31.60
3	1.64	2.35	3.18	4.54	5.84	7.45	12.92
4	1.53	2.13	2.78	3.75	4.60	5.60	8.61
5	1.48	2.02	2.57	3.37	4.03	4.77	6.87
6	1.44	1.94	2.45	3.14	3.71	4.32	5.96
7	1.42	1.89	2.37	3.00	3.50	4.03	5.41
8	1.40	1.86	2.31	2.90	3.36	3.83	5.04
9	1.38	1.83	2.26	2.82	3.25	3.69	4.78
10	1.37	1.81	2.23	2.76	3.17	3.58	4.59
11	1.36	1.80	2.20	2.72	3.11	3.50	4.44
12	1.36	1.78	2.18	2.68	3.06	3.43	4.32
13	1.35	1.77	2.16	2.65	3.01	3.37	4.22
14	1.35	1.76	2.14	2.62	2.98	3.33	4.14
15	1.34	1.75	2.13	2.60	2.95	3.29	4.07
∞	1.28	1.65	1.96	2.33	2.58	2.81	3.29

Further Reading

The following references are sources of more detailed information on the
techniques included in this book. The first four are of a general nature,
the remainder being specialist texts dealing with specific techniques or
groups of techniques.

1. FIFIELD, F.W. and KEALEY, D., Principles and Practice of Analytical
    Chemistry, 2nd edn. International Textbook Company, Glasgow and London,
    1983.

2. SKOOG, D.A. and WEST, D.M., Fundamentals of Analytical Chemistry,
    4th edn. CBS College Publishing, New York, 1982.

3. VOGEL, A.I., Textbook of Quantitative Inorganic Analysis, 4th edn.
    Longman, London,1978.

4. WILLARD, H.H., MERRITT, L.L.JR., DEAN, J.A. and SETTLE, F.A. JR.,
    Instrumental Methods of Analysis, 6th edn. Wadsworth Publishing Company,
    Belmont, California, 1981.

5. BAILEY, P.L., Analysis with Ion Selective Electrodes, 2nd edn. Heyden ,
    London, 1980.

6. EBDON, L., An Introduction to Atomic Absorption Spectrometry. Heyden,
    London, 1982.

7. BOLTZ, D.F. and HOWELL, J.A., Colorimetric Determination of Non-Metals,
    2nd edn. Wiley, New York, 1978.

8. FLASCHKA, H.A. and BARNARD, A.V., Quantitative Analytical Chemistry, 2nd
    edn. Willard Grant Press (Wadsworth), Belmont, California, 1981.

9. HAMILTON, R.J. and SEWELL, P.A., Introduction to High Performance Liquid
    Chromatography, 2nd edn. Chapman and Hall, London, 1982.

10. Handbook of Electrode Technology. Orion Research, Cambridge, Mass., 1982.

11. KEALEY, D., Infrared Spectrometry in Chemical Analysis, Interactive Software for the BBC B Microcomputer. Wiley, Chichester, UK, 1985.

12. PERRY, J.A., Introduction to Analytical Gas Chromatography, History, Principles and Practice (Chromatography Science Series, Vol 14). Marcel Dekker, New York, 1981.

13. POOLE, C.F. and SCHUETTE, S.A., Contemporary Practice of Chromatography. Elsevier, New York, 1984.

14. POUCHERT, C.J., The Aldrich Library of Infrared Spectra. Aldrich Chemical Co. Inc., 1970.

15. SANDELL, E.B. and ONISHI, H., Photometric Analysis of Trace Elements: General Aspects, 4th edn. Wiley, New York, 1978.

16. BRAITHWAITE, A. and SMITH, F., Chromatographic Methods, 4th edn. Chapman and Hall, London, 1985.

A.1 A Study of Acid–Base Titrations

Object

To plot titration curves from potentiometric data for strong acid/strong base, weak acid/strong base and salt/strong acid systems and to select suitable visual indicators based on the results.

Introduction and Theory

Volumetric (titrimetric) analysis in general is a rapid means of quantitative analysis that is capable of high precision (0.1 per cent or better) and accuracy, provided that Grade A volumetric glassware is used and care taken in the preparation of solutions and in the use of pipettes and burettes. All glassware should be cleaned before use (nitric/chromic acid) and thoroughly rinsed with distilled or deionized water.

Acid–base titrations are often performed routinely to monitor the acidity or alkalinity of solutions used in industrial processes, to establish the condition of a coolant, plating bath or a used product, or as a method of assay. Visual indicators must undergo colour changes over a pH range that includes that of the equivalence (neutralization) point otherwise significant errors will be incurred due to over or under titration. The pH of the equivalence-point can be established potentiometrically by measuring pH with a glass electrode as a function of volume of titrant added and plotting the data manually or automatically using an autotitrator. Routine titrations can be automated if the equivalence-point is to be detected potentiometrically. This is essential where samples are coloured or other conditions prevent the use of a visual indicator.

Requirements

Solutions of sodium hydroxide, hydrochloric acid, ethanoic (acetic) acid and sodium phenate (prepared from phenol and sodium hydroxide),each 0.1 mol dm^{-3}

Buffer tablets or powders, pH 4 and pH 7 or pH 9

A range of visual pH indicator solutions

pH meter, indicator (glass) and reference (calomel) electrodes

Magnetic stirrer

25-cm^3 burette, 10-cm^3 pipette, 100-cm^3 beakers

Procedure

1. Calibrate the pH meter using two standard buffer solutions (e.g. pH 4 and pH 9).

2. Pipette 10 cm^3 of 0.1 mol dm^{-3} hydrochloric acid into a 100-cm^3 beaker and place it on the magnetic stirrer. Rinse the electrodes with distilled or deionized water and immerse them in the solution, adding sufficient water to ensure that the bulb of the glass electrode is fully covered.

3. Add 0.1 mol dm^{-3} sodium hydroxide solution from the burette in 1.0 cm^3 amounts (0.1 cm^3 in the region of the equivalence-point) to a total of 20 cm^3. Record the pH meter reading after each addition, allowing it to stabilize between readings if necessary, stirring the solutions continuously.

4. Pipette 10 cm^3 of 0.1 mol dm^{-3} ethanoic acid into a 100-cm^3 beaker and repeat the procedure.

5. Pipette 10 cm^3 of 0.1 mol dm^{-3} sodium phenate into a 100-cm^3 beaker and titrate with 0.1 mol dm^{-3} HCl, recording the pH values as before.

6. Make separate 10-fold dilutions of the hydrochloric acid and sodium hydroxide solutions and repeat the first titration (2 and 3). Make further 10-fold dilutions and repeat the titration.

Results

1. Plot a pH v. volume of titrant curve for each titration and determine the equivalence-point volume and pH.

2. Select a suitable indicator for each of the titrations and check its performance by carrying out a rapid titration to a visual end-point and measuring the final pH.

Discussion

1. Write equations for each titration reaction and comment on the special case of sodium phenate.

2. Comment on the performance of each selected visual indicator and the end-point pH compared to that given by the potentiometric titration.

3. How do the concentrations of the titrant and titrand affect the choice of visual indicator?

4. Suggest how a mixture of ethanoic and hydrochloric acids might be analysed.

Instructor's Note

Capable students should be able to complete all the titrations within a 4-hour practical period. For shorter periods, or with less experienced students, they could be asked to perform selected titrations only. The dilution effect could also be studied for the ethanoic acid and sodium phenate titrations.

A.2   The   Identification of an Unknown Weak Acid by Potentiometric Titration

Object

To determine the acid dissociation constant(s) and the relative molecular mass of an unknown weak acid from potentiometric titration data and hence to identify the acid.

Introduction and Theory

The dissociation constant, K_a, of a weak monoprotic acid, HA, is given by the expression

$$K_a = \frac{[H^+][A^-]}{[HA]} \text{ or } pK_a = pH + \log_{10}\frac{[HA]}{[A^-]}$$

where A^- is the anion of the acid and [ ] denotes concentrations (strictly activities). Similar expressions can be written for the first and subsequent dissociations of a polyprotic acid. At the half-neutralization point in the titration of a monoprotic acid with sodium hydroxide, $[HA] = [A^-]$, and an approximate value for the pK_a of the acid is therefore given by the pH of the solution at this stage. For a diprotic acid, an approximate value for the second dissociation constant, pK_2, can be calculated using the expression

$$pK_2 = 2pH - pK_1$$

Where pK_1 is the first dissociation content and the pH is that of the solution at the first equivalence-point. The total titre allows the relative molecular mass (RMM) to be calculated, provided a known amount of the acid has been titrated. The values for the dissociation constant(s) and the RMM can be used to identify the acid.

Potentiometry involves measuring the potential of an electrochemical cell consisting of the analyte solution, an indicator electrode, e.g. a glass electrode sensitive to solution pH, and a reference electrode having a stable potential independent of solution composition. The pH changes that occur during an acid-base titration can be monitored using a calibrated pH meter, and a titration curve of pH against volume of alkali added plotted manually or automatically by an autotitrator. The curve generated by this 'potentiometric titration' can be used to determine the equivalence point(s) and hence to provide additional information about the acid.

Requirements

Standard sodium hydroxide solution, 0.1 mol dm^{-3}

25-cm^3 burette or microburette

250-cm^3 volumetric flask and 50-cm^3 pipette

pH meter, indicator (glass) and reference (calomel) electrodes

Buffer tablets or powders, pH 4 and pH 7 or pH 9

Magnetic stirrer

Unknown weak mono- or diprotic acid (See Instructor's Note 1)

Procedure

1. Weigh accurately 0.25 to 0.35 g of the unknown acid into a 250-cm³ beaker, add about 175 cm³ of distilled or deionized water, cover with a watch glass and warm to no more than 40°C.   Maintain at this temperature and stir occasionally until the acid has completely dissolved.   Cool, transfer quantitatively to a 250-cm³ volumetric flask and dilute to the mark.

2. Calibrate the pH meter using the buffer tablets or powders, checking that the meter or readout is correct at pH 4 <u>and</u> at pH 7 or pH 9.

3. Pipette a 50 cm³ aliquot of the sample solution into a 150-cm³ beaker, immerse the glass and reference electrodes in the solution and stir gently and continuously with the magnetic stirrer.

4. Record the pH of the sample solution, add a 0.5-cm³ increment of the standard sodium hydroxide solution from a burette and note the new pH reading.   Repeat the procedure adding 0.5-cm³ increments (0.1 cm³ near the equivalence-point when the pH reading changes more rapidly) until 10 cm³ have been added.   The final pH should be in the region of 10 to 12. Titrate a further 50-cm³ aliquot of the sample solution in the same manner, but adding larger increments before and after the equivalence-point region to save time.

Results

1. For each titration, plot a graph of pH against the volume of standard sodium hydroxide solution added.   Determine the equivalence-point volume from the point of inflection (the second in the case of a diprotic acid). Calculate the mean RMM of the acid.

2. Calculate the approximate mean pK_a value(s) of the acid using the equation(s) given in the Introduction and Theory section.   Identify the acid from the calculated pK_a and RMM values and published data.

Discussion

1. Explain why the calculated pK_a values are only approximate.   Is the RMM value also an approximation?

2. Could visual acid-base indicators be used in this experiment?

3. In general, what other techniques are used for the determination of pK_a values?

4. Give examples of other types of titration for which the equivalence-point can be detected potentiometrically.

EXPERIMENT A.2

Instructor's Notes

1. There are a number of suitable mono and diprotic acids for this experiment, e.g. benzoic, boric, maleic, malonic, diethyl malonic, mandelic and sulphanilic.

2. The determination of pK_a can be made more rigorous by calculating a series of values using the expression

$$pK_a = pH + \log_{10} \frac{[Y] - [NaOH] + [OH^-] - [H^+]}{[NaOH] - [OH^-] + [H+]}$$

where [Y] is the total concentration of the unknown acid. This allows pK_a values at different stages of the titration to be calculated from which an overall mean can be derived. (Only H^+ and OH^- activities are taken into account.)

3. Procedure 4 and Results 1 will require some modification if an autotitrator is used.

A.3   The Identification of an Unknown Oxirane (epoxide) by Non-Aqueous Titrimetry

Object

To determine the relative molecular mass of an unknown oxirane (epoxide) by non-aqueous titration with hydrogen bromide in ethanoic (acetic) acid, to test for the presence of elements other than C, H and O and hence to identify the oxirane (epoxide).

Introduction and Theory

Volumetric (titrimetric) analysis in general is a rapid means of quantitative analysis that is capable of high precision (0.1% or better) and accuracy provided that Grade A volumetric glassware is used and care taken in the preparation of solutions and in the use of pipettes and burettes. All glassware should be cleaned before use (nitric/chromic acid) and thoroughly rinsed with distilled or deionized water.

Non-aqueous titrations are a valuable extension to aqueous titrimetry when the analysis of water-insoluble materials is required or where the provision of a non-aqueous environment can enhance the acidic or basic nature of substances that are extremely weak acids or bases in the presence of water.

Compounds containing the oxirane ring can be determined quantitatively by titration with a solution of hydrogen bromide in glacial ethanoic (acetic) acid, the hydrogen bromide cleaving the ring and brominating the compound. The end-point of the titration is indicated visually using crystal violet. The ethanoic (acetic) acid acts both as a solvent for the oxirane compound and to enhance the acidity of the hydrogen bromide.

Requirements

Sodium carbonate, AnalaR grade

Glacial ethanoic (acetic) acid, GPR grade

Hydrogen bromide/ethanoic acid reagent, 45–50%

Crystal violet indicator solution, 0.1% w/v in glacial ethanoic acid

Unknown oxirane (epoxide)   (see Instructor's Note 1)

Elemental sodium and reagents for sodium-fusion tests

100-cm^3 volumetric flasks, 5-cm^3 pipette, burette

Beakers and conical flasks

Procedure

1. Dry about 1.5 g of sodium carbonate for 2 hours at 150°C and cool in a desiccator. Accurately weigh about 1 g into a dry 100-cm^3 beaker and dissolve it in a small volume of glacial ethanoic acid, covering the beaker with a watch glass. Transfer the solution quantitatively to a dry 100-cm^3 volumetric flask and dilute to the mark with glacial ethanoic acid.

    CAUTION – Handle glacial ethanoic acid with care and avoid breathing the fumes.

2. Dilute 2 cm^3 of the hydrogen bromide/ethanoic acid reagent to 100 cm^3 with glacial ethanoic acid in a dry flask. Titrate 5 cm^3 aliquots of the standard solution prepared in 1 (diluted with a little glacial ethanoic acid) with the diluted hydrogen bromide/ethanoic acid solution using about 5 drops of the crystal violet solution as indicator. The colour change is violet → blue → emerald green.

3. Weigh about 150 mg of the unknown oxirane compound into a small dry conical flask, add several cm^3 of glacial ethanoic acid and titrate with the diluted hydrogen bromide/ethanoic acid solution as in 2. Repeat at least once.

4. Check the oxirane for elements other than C, H and O by sodium fusion followed by the appropriate chemical tests.

Results

1. Calculate the molarity of the hydrogen bromide/ethanoic acid solution.

2. Calculate the mean relative molecular mass of the oxirane.

3. Identify any element other than C, H and O.

4. Identify the oxirane and suggest the most likely structural formula.

Discussion

1. Write equations for the following reactions:

    (a) sodium carbonate and glacial ethanoic acid
    (b) the standardization titration
    (c) the oxirane titration

2. Suggest a simple spot-test for oxiranes based on this determination.

3. What other compounds might also react with hydrogen bromide in a similar manner to oxiranes?

4. Suggest an alternative means of detecting the end-point in this type of titration.

Instructor's Notes

1. Low molecular mass alkyl or haloalkyl substituted oxiranes (epoxides) are suitable for this determination. Most are liquids.

2. Students should be reminded to handle glacial ethanoic acid with great care, due to its toxicity, preferably dispensing it in a fume cupboard.

3. Titrations should be continued until the emerald-green end-point colour persists for at least one minute. Any water in the system will inhibit a sharp colour change.

A.4 The Identification of a Metal Salt by Complexometric Titration

Object

To determine the relative molecular mass of an unknown metal salt by titration with ethylenediaminetetraacetic acid (EDTA), to establish the identity of the anion by chemical tests and hence to identify the metal salt.

Introduction and Theory

Volumetric (titrimetric) analysis in general is a rapid means of quantitative analysis that is capable of high precision (0.1 per cent or better) and accuracy provided that Grade A volumetric glassware is used and care taken in the preparation of solutions and in the use of pipettes and burettes. All glassware should be cleaned before use (nitric/chromic acid) and thoroughly rinsed with distilled or deionized water.

Many metals in solution can be determined by titration with a standard solution of a complexing agent of which EDTA is the most widely used. The reaction with most metals has a 1:1 stoichiometry and is quantitative at pH values determined by the formation constant of the complex. The reagent, which can be represented as H_4Y, is normally used in the form of its disodium salt, H_2Na_2Y, as it dissolves readily in water, the undissociated acid being much less soluble. Metallochromic indicators, such as xylenol orange, form coloured complexes with metal ions but change colour when the metals are fully complexed with EDTA as at the end-point in a titration. As these indicators are also sensitive to pH changes, solutions to be titrated must be well buffered.

Requirements

Ethylenediaminetetraacetic acid, disodium salt, AnalaR grade

Xylenol orange indicator, 0.5%w/v aqueous solution

Hexamine (hexamethylenetetramine), GPR grade

Dilute nitric acid, 0.1M

Unknown metal salt (see Instructor's Note 1)

Burette, pipette (25-cm^3), 250-cm^3 volumetric flasks (2), 250-cm^3 conical flask

Test-tubes and reagents for qualitative tests for anions

Procedure

1. Prepare a 0.05 mol dm^{-3} solution of EDTA by dissolving the appropriate amount in about 100 cm^3 of distilled or deionized water with warming. Cool, transfer the solution quantitatively to a 250-cm^3 volumetric flask and dilute to the mark.

2. Prepare a solution of the unknown metal salt by dissolving between 3 g and 4 g (accurately weighed) in about 100 cm^3 of distilled or deionized water with warming. Cool, transfer the solution quantitatively to a 250-cm^3 volumetric flask and dilute to the mark.

    CAUTION: The unknown salt may be toxic; handle with care; use a pipette filler.

3. Pipette 25 cm^3 of the salt solution into a 250-cm^3 conical flask, dilute to about 50 cm^3 with distilled or deionized water and add 2 to 4 drops of xylenol orange indicator. If the solution is red to reddish-purple, add 0.1M nitric acid <u>dropwise</u> until it <u>just</u> turns yellow. Add about 2 g of solid hexamine to buffer the solution to pH 6 to 6.5. The colour should revert to a red to reddish-purple at this stage.

4. Titrate the solution with 0.05 mol dm^{-3} EDTA to a red or reddish-purple → yellow end-point. If a precipitate forms initially, it should re-dissolve near the end-point.
    Repeat the titration until titres agree to within 0.5 per cent.

5. Test the salt for anions using standard chemical tests and confirm the identity of the metal after a tentative identification has been made (see Results).

Results

1. Using the titration data, calculate the relative molecular mass (RMM) of the unknown salt.

2. Deduce the identity of the metal and confirm it by a chemical test (procedure 5). It may be necessary to take account of water of crystallization in a hydrated salt.

Discussion

1. Why does EDTA form such strong complexes with many metals?

2. Why is it necessary to buffer the metal salt solution at pH 6 to 6.5 before titrating with EDTA?

3. Suggest two alternative methods of end-point detection in complexometric titrations.

4. What other analytical techniques could be used to identify an unknown metal salt?

Instructor's Notes

1. Suitable metal salts include those of cadmium, zinc, lead and the lanthanide elements. If the salt is a hydrate, the number of water molecules must be quoted or established by weight loss on drying.

2. Less experienced students will need guidance in the selection of suitable chemical tests for anions and the metal.

A.5 The Determination of Magnesium and Zinc in a Mixture and/or the Determination of Zinc in Pharmaceutical Preparations by Ion Exchange Separation and Complexometric Titration

Object

To separate a mixture of magnesium and zinc by anion exchange and to determine each metal quantitatively by titration with ethylenediaminetetraacetic acid (EDTA) and/or to isolate and determine zinc in vitamin/mineral formulations by anion exchange and EDTA titration.

Introduction and Theory

Ion exchange resins are insoluble organic polymers in bead form to which ionizable acidic or basic groups have been chemically bonded. In the presence of aqueous solutions, the bonded groups are ionized, and one of the ions of each group (the counter ion) can exchange with chemically different ions of like charge present in the solution. When packed into a column, ion exchange resin beads can selectively and quantitatively remove ions of a particular element or compound from a sample solution that is allowed to percolate slowly through the resin bed. The retained ions can subsequently be removed (eluted) by the passage of another solution through the column. This procedure results in the complete separation of certain of the elements or compounds that were present in the original solution. Ion exchange separations are used as preliminary stages in analytical procedures when one component of a sample interferes with the determination of another. The resins are also used to concentrate trace amounts from very dilute samples.

Many metals in solution can be determined by titration with a standard solution of a complexing agent of which EDTA is the most widely used. The reactions with magnesium and zinc have 1:1 stoichiometries and are quantitative above pH 9 and pH 3 respectively. EDTA, which can be represented as H_4Y, is normally used in the form of its disodium salt, H_2Na_2Y, as it dissolves readily in water, the undissociated acid being much less soluble. Metallochromic indicators such as methylthymol blue and eriochrome black T form coloured complexes with metal ions but change colour when the metals are completely complexed with EDTA as at the end-point in a titration. As these indicators are also sensitive to pH changes, solutions to be titrated must be well buffered.

Requirements

Ethylenediaminetetraacetic acia, disodium salt, AnalaR grade

Eriochrome black T indicator solution, 1% w/v in 3:1 triethanolamine/ethanol

Methylthymol blue indicator, 1:100 w/w with AnalaR KNO_3

Hexamine (hexamethylenetetramine), GPR grade

Anion exchange resin, e.g. Zerolit FF Cl$^-$ form (50-100 or 100-200 mesh, analytical or chromatographic grade)

Glass ion-exchange column, about 20 cm x 0.6 cm

Ammonia-ammonium chloride buffer solution, pH 10
(142 cm^3 of concentrated aqueous ammonia, sp. gr. 0.880, and 17.5g of AnalaR grade ammonium chloride diluted to 250 cm^3 with distilled or deionized water)

Concentrated hydrochloric acid, AnalaR grade

Dilute hydrochloric acid, 4M and 1.5M

Dilute nitric acid, 0.5M

Ammonium hydroxide solution, 4M

Burette, pipette ($10-cm^3$), conical flasks ($250-cm^3$), beakers ($30-cm^3$), measuring cylinders ($25-cm^3$)

Magnesium and zinc solutions of unknown concentration

Proprietary vitamin/mineral formulation containing zinc (see Instructor's Note 1)

Potassium cyanide, AnalaR - CAUTION, TOXIC, HANDLE WITH CARE

Procedure

1. Prepare a 0.01 mol dm^{-3} solution of EDTA by dissolving the appropriate amount in about 100 cm^3 of distilled or deionized water with warming. Cool, transfer the solution quantitatively to a $250-cm^3$ volumetric flask and dilute to the mark.

2. Prepare the anion exchange resin by mixing about 10 g with 200 cm^3 of water in a $250-cm^3$ beaker. Stir well and decant off most of the water to remove fine particles of resin. Repeat the process twice more, then transfer the resin to the column, ensuring that the resin bed remains covered with water when it has settled. Pass 50 cm^3 of 1.5M HCl through the column to condition it. <u>N.B.</u> Do not allow the top of the resin bed to become dry.

3. For vitamin/mineral formulations only:

   To a capsule/tablet* of the formulation in each of two $30-cm^3$ beakers add 5 cm^3 of 1:1 HCl/water. Cover the beakers with watch glasses and heat them on a water bath until decomposition is complete. If residues remain, filter the hot solutions through moistened medium-porosity filter papers (e.g. No 40 or 540) using small filter funnels or Hirsch funnels to which suction can be applied. Collect the filtrates in or transfer them to two $25-cm^3$ measuring cylinders. Rinse the beakers and filter papers with several small portions of distilled or deionized water adding the washings to the respective sample solutions until the volume in each cylinder is 20 cm^3.

4. For magnesium/zinc mixtures only:

   Pipette 10 cm^3 of the magnesium and zinc solution into a $30-cm^3$ beaker and add 6 cm^3 of 4M HCl.

5. Pass one of the capsule/tablet solutions prepared in 3 or the magnesium zinc solution prepared in 4 down the ion exchange column at not more than 3 cm^3/minute, collecting the eluent in a $250-cm^3$ conical flask. Rinse the beaker with 5 x 10 cm^3 portions of 1.5M HCl passing each down the column and collecting the eluent in the flask.
   The zinc is retained on the column. For vitamin/mineral formulations, discard the contents of the flask and continue at paragraph 7.

6. Add 20 cm^3 4M NH_4OH solution to the contents of the conical flask, which contains the magnesium, followed by 10 cm^3 of the ammonia-ammonium chloride buffer solution and about 0.2 g of KCN. Ensure that the KCN has dissolved then add sufficient eriochrome black T indicator to produce a pale-pink coloration in the solution. Titrate the magnesium with 0.01 mol dm^{-3} EDTA to a pink → violet → pure blue end-point.

7. Pass 120 cm^3 of 0.5 HNO_3 down the column at not more than 5 cm^3/minute to elute the zinc, collecting the eluent in a conical flask. Add 15 cm^3 of 4M NaOH solution followed by about 2 g of hexamine. Adjust the pH of the solution to between 5.8 and 6.5 by adding more sodium hydroxide dropwise,

---

* More than one may be required if the zinc content is less than about 2 mg.

then add enough methylthymol blue indicator to produce a distinct violet-blue coloration in the solution. Titrate the zinc with 0.01 mol dm^{-3} EDTA to a violet-blue → yellow end-point.

8. Recondition the ion exchange column by passing 100 cm^3 of 4M HCl through it followed by 20 cm^3 of 1.5M HCl then repeat the procedure starting at 4 or 5 as appropriate.

Results

1. <u>For vitamin/mineral formulations only</u>:

    Calculate the amount of zinc in mg in each capsule/tablet and the mean of the two.

2. <u>For magnesium/zinc mixtures only</u>:

    Calculate, from the mean titres for magnesium and zinc, the concentration of each metal in mg cm^{-3} of the original magnesium and zinc solution.

Discussion

1. Explain why the ion exchange resin retains the zinc but not the magnesium under the conditions of the experiment. Why is dilute nitric acid used to elute the zinc?

2. If a vitamin/mineral formulation has been analysed, comment on the behaviour of other metals present in the product during the ion exchange step. If the magnesium/zinc mixture has been analysed, comment on the use of two different indicators.

3. Suggest a masking agent and a procedure that could be used to enable both metals to be determined without a prior separation.

4. What other techniques are suitable for the determination of magnesium and zinc in a mixture? What are their advantages/disadvantages?

Instructor's Notes

1. There are a number of formulations available from chemists (drug stores) and health-food shops that contain sufficient zinc (and numerous other metals) for this analysis.

2. Preparation of the ion exchange resin column and reagent solutions beforehand is advisable for less experienced students.

3. The magnesium and zinc contents of the sample solution should be in the region of 0.2 mg to 0.7 mg per cm^3 and contain 1 drop of concentrated HCl to prevent hydrolysis.

4. A statistical analysis of class or group results is a useful exercise.

A.6 The Gravimetric Determination of Nickel in Steel

Object

To determine the nickel content of a steel by precipitation and weighing as the dimethylglyoxime complex.

Introduction and Theory

Gravimetric analysis provides a highly accurate and precise means (0.1 per cent or better) of determining the major and minor constituents (predominantly metals) of samples such as alloys, ores and minerals. It is sometimes used to establish the reliability of alternative instrumental methods that have now largely replaced gravimetry for routine analyses. It involves the isolation of an analyte in a weighable form of known stoichiometry. The most common procedure is to precipitate the analyte from solution in the form of a sparingly soluble compound formed by the addition of a suitable reagent. The precipitate is isolated by filtration, washed to remove co-precipitated impurities, then dried and weighed. In some cases, ignition to an oxide provides a weighable form of more reliable stoichiometry.

Requirements

Dimethylglyoxime, AnalaR grade, 1% w/v solution in ethanol

Concentrated nitric and hydrochloric acids, AnalaR grades

Citric acid, AnalaR grade

Sintered-glass crucibles, medium porosity

400-cm^3 beakers

Steel sample (at least 1% nickel)

Procedure

1. Weigh accurately two 0.5 to 1 g* samples of the nickel - containing steel into 400-cm^3 beakers.   Add 15 cm^3 of concentrated HCl and 2 cm^3 of concentrated HNO$_3$ to each, cover with watch glasses and heat to boiling. Maintain the temperature near boiling until the samples have dissolved (ignore slight residues of silica at this stage).   Cool, dilute to about 50 cm^3 with distilled or deionized water and filter if necessary.   Wash the filter paper with small portions of hot water, adding the washings to the sample solutions, and dilute each to a final volume of about 250 cm^3.

2. Add 5 g of citric acid to each solution, neutralize with dilute aqueous ammonia then make just acid (pH 4 to 5) with dilute HCl.   Warm the solutions to about 70°C, add 25 cm^3 of the 1% dimethylglyoxime solution followed by dilute aqueous ammonia dropwise, with stirring, until they are slightly ammoniacal.   Stir well, allow to stand on a steam-bath for 30 minutes and then at room temperature until cold (about 1 hour).

* Depending on the expected nickel content of the steel.

3. Filter the solutions through previously dried and weighed sintered-glass
    crucibles.   Test the filtrates for completeness of precipitation by adding
    a few drops of the dimethylglyoxime solution; if more nickel complex forms,
    return the filtrate to the beakers, add 5 cm^3 more reagent and repeat the
    procedure commencing with digestion on the steam-bath.   Finally, wash with
    cold distilled or deionized water until the washings are chloride-free.
    Dry at 120°C for one hour, cool in a desiccator and weigh as $Ni(C_4H_7O_2N_2)_2$.

Results

1. Calculate the per cent nickel in each sample and the mean of the two.

2. If results from other students are available, calculate
    (a) the overall mean
    (b) the estimated standard deviation and relative precision (p.17)
    (c) the confidence interval about the mean (p.17).

Discussion

1. What is the structure of the nickel dimethylglyoxime complex and with what
    other metals does this reagent react?

2. What are the roles of the nitric and citric acids?

3. Why are samples digested and allowed to stand before filtration?

4. Suggest alternative methods for the determination of nickel in steel and
    comment on their potential advantages/disadvantages.

Instructor's Notes

1. The properties of the nickel dimethylglyoxime complex are not ideal in that
    the precipitate tends to 'creep' during filtration.   Students should
    exercise particular care during this step so as to avoid losses.

2. Drying the precipitate at 160°C may be necessary where students have added
    a considerable excess of the reagent.   The excess reagent will be
    volatilized at this temperature.

A.7 Adsorption Thin–Layer Chromatography

Object

To illustrate the separation of compounds of differing polarities using adsorption TLC and various methods of visualization.

Introduction and Theory

Thin-layer chromatography (TLC) is a sensitive, inexpensive and simple technique used for the separation and characterization of the components of mixtures. Its applications include the investigation of purity, following the course of an organic synthesis, and screening tests. It is of particular value in the analysis of foodstuffs, drugs, plant extracts and dyestuffs where it provides a rapid means of detecting contaminants and adulterants.

Chromatographic methods exploit the differing affinities substances may have for a mobile and stationary phase which results in differential rates of migration. Most TLC separations involve silica gel plates where the rate of migration is determined by the relative polarities of the substance and the mobile phase, the mechanism being primarily one of surface adsorption. Separated components are visualized by spraying the plate with a chromogenic reagent, exposing it to a vaporized reagent or using a plate that has been impregnated with a fluorescent indicator during manufacture. Components are characterized by their R_f value, the distance travelled from the point of application relative to the distance travelled by the mobile phase (solvent) front (p. 10).

Requirements

Chloroform solutions of: squalene, α-tocopherol (vitamin E), cholesterol, methyl oleate and methyl stearate, 2 to 5% w/v, GPR grades

Chloroform solutions of mixtures of three of the above compounds

Phosphomolybdic acid spray reagent, 12% w/v in ethanol, GPR grades

Iodine, GPR grade

Diethylether, methylbenzene (toluene), GPR grade

Chromatographic development tanks or tall beakers

Plastic or aluminium foil backed silica gel TLC plates (some plates should incorporate a fluorescent indicator)

Large sheets of absorbent paper

Melting-point tubes

Pressurized spray-guns

Infrared and ultraviolet lamps

Hot–air blower

Procedure

1. Line a development tank with absorbent paper for about two–thirds of its height and add sufficient of a mixture of 5% diethylether/95% methylbenzene to soak the paper and cover the bottom of the tank to a depth of about 0.5 cm.   Cover the tank and allow it to stand for at least 15 minutes.

2. Prepare a number of sample applicators from melting-point tubes by softening the middle in a micro-burner flame and drawing the tube out to create a very narrow constriction in the centre.   Snap each tube in half when cold to produce two applicators with tapered ends.

3. Select a TLC plate with or without fluorescent indicator added (see 6 below).   Draw a very faint pencil line across the plate about 1 cm from one edge and parallel to it and apply spots of each compound or mixture at 1-cm intervals along the line.
    N.B. Use each applicator ONCE ONLY then discard it so as to minimize the risk of contamination.

4. Place the plate into the tank with the sample spots at the bottom, but not covered by the developing solvent, replace the cover and allow the solvent front to travel up the plate for about two-thirds to three-quarters of its length (~40 minutes).

5. Remove the plate, mark with a pencil the point to which the solvent front has reached and allow it to dry.   (Gentle heating with a hot-air blower will help.)

6. Visualize the compounds by one of the following three methods:

    (i) Phosphomolybdic acid spray (FUME CUPBOARD)

        Spray the plate lightly and evenly and heat strongly under an IR lamp for 5 to 10 minutes or until the spots are fully visualized.

    (ii) Iodine (FUME CUPBOARD)

        Place the plate in a tall beaker or dry chromatographic development tank containing a few crystals of iodine.   Cover the top to prevent the escape of iodine vapour and heat gently with an IR lamp or hot-air blower until the spots are fully visualized.

    (iii) Fluorescence

        View the plate under a shielded uv lamp and mark the positions of the observed spots.

7. Compare the results (a) from at least two different visualizing methods and (b) by applying different-sized samples to the TLC plates.

Results

1. Measure the R_f value of each visualized spot and tabulate the results for the various plates developed.

2. Identify the components of each mixture.

Discussion

1. Comment on the relative polarities of the five compounds.

2. Comment on the different methods of visualization used and the chemical reactions involved.   Are all the compounds visualized with each method?

3. What is the observed effect of sample size on the visualized spots?   What conclusions can be drawn?

4. a. How sensitive a technique is TLC and what are the major two factors affecting it?

    b. What experimental factors affect R_f values?

Instructor's Notes

1. Students should appreciate the need to work on a clean surface, to handle TLC plates with care and to understand how easily the samples can become cross-contaminated through carelessness.

2. Contact between the edges of the plate and the solvent-soaked absorbent paper lining the tank should be avoided as uneven movement of the solvent front may result.

3. Most students can prepare, develop and visualize at least two plates during a 3-hour practical session.

4. An alternative visualizing reagent is 50% sulphuric acid, with which all organic compounds can be detected. Glass-backed plates are required as strong heating (> 150°C) is necessary to visualize the spots.

A.8 The Potentiometric Determination of Fluoride in Tap Water

Object

To study the effect of pH on the response of a fluoride ion-selective electrode and to determine the fluoride content of a sample of tap water.

Introduction and Theory

Ion-selective electrodes enable the concentrations of such inorganic anionic species as halides, nitrate, nitrite and cyanide, some metal ions and dissolved gases to be measured and monitored over an extremely wide range extending from sub-ppm levels to over 1 mol dm^{-3} solutions. They find applications throughout industry and in other institutions in the analysis of surface waters, soils, biological fluids, foodstuffs, beverages and other manufactured products.

The level of naturally occurring or added fluoride in drinking water is of interest because of its relation to the incidence of dental caries. The fluoride electrode used in this experiment is based upon a single crystal membrane of lanthanum fluoride. This membrane is sensitive to the presence of fluoride ions in a sample solution, and a membrane potential is developed that is inversely proportional to the concentration of fluoride ions in the solution. The relation is given by a form of the Nernst equation (p. 9).

$$E_{cell} = k - 0.059 \log_{10}[F^-]$$

where E_{cell} is the cell potential as measured on a millivolt or pH meter, k is a constant and $[F^-]$ denotes the concentration of fluoride ions in the sample solution. The value of E_{cell} changes by 59 millivolts (approximately) per tenfold change in fluoride concentration. Individual electrodes may vary by several millivolts from this value.

The response of the electrode is, however, pH-dependent, and for accurate measurements a buffer solution is needed to control the pH within the region of optimum response.

Requirements

500 ppm F^- stock solution (1.105 g of AnalaR NaF in 1 dm^3 of distilled or deionized water)

Phosphate buffer solution, pH 5.5 (4.4 g of AnalaR $Na_2HPO_4.12H_2O$, 13.7 g of AnalaR $NaH_2PO_4.2H_2O$ and 74 g of Analar KCl in 1 dm^3 of distilled or deionized water)

Hydrochloric acid, 4M

Sodium hydroxide solution, 4M

Buffer tablets or powders, pH 4 and pH 7 or pH 9

Tap water sample

pH meter

Expanded scale millivolt/pH meter

Fluoride ion selective electrode and reference electrode

Glass electrode (pH) and reference electrode (calomel)

250-cm^3 volumetric flask and graduated pipette

Plastic or glass beakers

Magnetic stirrer

Procedure

1. Prepare a 5 ppm fluoride solution by pipetting an aliquot of the stock fluoride solution into a 250-cm^3 volumetric flask, adding 125 cm^3 of the phosphate buffer solution and diluting to the mark with distilled or deionized water.

2. Calibrate the pH meter with suitable buffer solutions (pH 4 and pH 7 or pH 9). Investigate the response of the fluoride electrode (use the meter on the expanded mV range) as a function of pH over the range 5.5 to 11 at the 5 ppm level using about 100 cm^3 of the solution prepared in 1. Make pH adjustments by adding 4M sodium hydroxide dropwise as required and stir the solution continuously (six or seven pairs of readings are sufficient).

   N.B. The response of the fluoride electrode is sluggish at high and low pH values; take readings after 2 to 3 minutes if they are unstable.

3. Rinse the electrodes thoroughly and immerse them for several minutes in some of the phosphate buffer solution. Repeat the investigation over the pH range 5.5 to 1 using a fresh 100-cm^3 portion of the 5 ppm fluoride solution prepared in 1. Make pH adjustments by adding 4M HCl dropwise as required and stir the solution continuously (six or seven pairs of readings are sufficient).

4. Rinse the electrodes thoroughly and immerse them for several minutes in some of the phosphate buffer solution. Mix 50 cm^3 of the phosphate buffer solution with 50 cm^3 of tap water and measure the fluoride electrode response. Prepare two standards in 50% v/v phosphate buffer solution so as to 'bracket' the estimated sample concentration (see Results 2) and determine their electrode response.

Results

1. Plot a graph of fluoride electrode response (mV) as a function of pH.

2. Make a preliminary estimate of the fluoride content of tap water by using the data from 2 and 3 and assuming that the electrode has a Nernstian response. Obtain a more accurate value using the data from the two 'bracketing' standards.

Discussion

1. Explain the shape of the curve of fluoride electrode response as a function of pH. How will the curve vary with F$^-$ concentration?

2. Why is it necessary to prepare a freshly diluted standard from a relatively concentrated stock solution?

3. How would polyvalent metal ions such as Al(III), Fe(III), La(III) and Ca(II) affect the fluoride electrode response?

4. Suggest a titrimetric procedure for the determination of fluoride.

Instructor's Notes

1. The 500 ppm stock fluoride solution should be stored in a plastic bottle.

2. A total ionic strength adjustment buffer (TISAB) may be used in place of
    the phosphate buffer (available commercially through BDH Ltd,) or can be
    easily prepared in the laboratory (see A Textbook of Quantitative Inorganic
    Analysis, 4th edn., A.I.Vogel,Longman, 1978)

3. Two or three cycle semi-log graph paper is ideal for plotting the data.   A
    line of slope 59 mV/decade change of concentration should be drawn through
    the 5-ppm point to provide the preliminary estimate of the fluoride level
    in the tap water sample.

4. It is recommended that a thermally insulating pad be placed between the
    sample solutions and the top of the magnetic stirrer as the fluoride
    electrode response is sensitive to temperature changes.   The motors of
    most stirrers become quite warm during prolonged use.

B.1 A Statistical Evaluation of Spectrometric Absorption Data including Linear Regression Analysis

Object

To estimate the precision and accuracy of visible spectrometric absorption data by statistical methods using solutions of cobalt (II) nitrate, and to compare the performance of several laboratory instruments.

Introduction and Theory

All measurements are subject to random (indeterminate) errors that are generally small in magnitude but introduce a degree of uncertainty into the measuring process. Such measurements can be treated statistically so as to provide a means of assessing the quality of the data in terms of its reproducibility, i.e. precision, and its accuracy, i.e. its closeness to a true or accepted value. The statistical treatment of spectrometric absorption data provides a good illustration of the use of statistical methods in analytical chemistry.

The random errors associated with absorbance measurements arise largely from electrical noise generated by instrumental components and circuitry, and from cell positioning uncertainty. The performance of spectrometers varies with their design, age and condition. It is, therefore, instructive to compare two or more instruments of different type or age with respect to the precision of a series of absorbance measurements made in an identical fashion, and to assess the accuracy of measuring the analyte concentration in an unknown solution.

For species in solution, the absorbance at a particular wavelength is related to concentration and path length by the Beer-Lambert law (see also p. 9).

$$A = \epsilon.b.c$$

where A is absorbance, c is the concentration in mol dm^{-3} , b is the path length in cm and ϵ is the molar absorptivity. A plot of A versus c for a series of standards over a given concentration range is called a calibration curve. It is ideally linear and should pass through the origin. Calibration curves are extensively used to establish instrument response before a sample of unknown concentration can be analysed. The precision of the calibration data, which can be determined by linear regression analysis (p.15), and the precision and accuracy of the measurement of an unknown concentration all need to be established.

Requirements

Cobalt (II) nitrate stock solution, 0.188 mol dm^{-3}

Cobalt (II) nitrate solution, unknown concentration

Burette and 50-cm^3 volumetric flasks

A selection of visible range spectrometers, preferably including a filter photometer and both single and double-beam spectrophotometers

1-cm path length glass cells (cuvettes)

Procedure

1. Prepare a series of nine cobalt standards by dispensing between 5 cm^3 and 45 cm^3 of the cobalt (II) nitrate stock solution into 50-cm^3 volumetric flasks from a burette, and diluting each to the mark with distilled or deionized water.   Fill a tenth flask with the undiluted stock solution.

2. (a) For a filter photometer, select a filter with a maximum transmission at or close to 510 nm.

    (b) For a single-beam spectrophotometer, check the absorbance of one of the more concentrated standards in the region of 510 nm and select the wavelength at which the absorbance is a maximum.

    (c) For a double-beam spectrophotometer, record the spectra of all the solutions between 450 nm and 550 nm.

    Measure all absorbances relative to a distilled or deionized water blank.

3. For the selected instrument, measure the absorbances of the standards and the unknown cobalt solution in the following sequence: std 1, unknown, std 2, unknown, std 3, unknown, std 4 ..... etc. until each standard has been measured once and the unknown ten times.   Ensure that the optical surfaces of the cells remain clean and dry.

4. Repeat the procedure for at least one more instrument of a different type.

Results

1. For each instrument, enter the concentrations and absorbances of the standards and the ten absorbance readings for the unknown in the appropriate columns of a copy of the data sheet (see p.51).   Perform a linear regresion analysis of the data (see p.15) as described in 2.

2. Complete the other columns on the sheet and calculate the following using the formulae given:

    molar absorptivity, ϵ

    calibration curve intercept, a

    estimated standard deviation about the calibration regression line, s_1

    mean concentration of the unknown cobalt solution, $\bar{c}_i$

    estimated standard deviation for the concentration of the unknown cobalt

    solution, s_2

    confidence limits about $\bar{c}_i$ based on s_2

    (see also pp. 15 to 18).

3. Plot the calibration data and draw the regression line using the appropriate value for a and the slope ϵ/b (where b, the path length of the cells used, is 1 cm).

    If a suitable computer program is available, use it to check the statistical calculations.

Discussion

1. What instrumental factors may cause apparent deviations from the Beer-Lambert law? Are they relevant in this experiment?

2. In general, how does the precision of absorbance measurements vary with the magnitude of the measured absorbance?

3. Comment on the comparative performances of the instruments used with respect to precision and accuracy and account for any differences. In general, what other major sources of error may arise in making measurements?

4. Calibration curves sometimes do not pass through the origin and may have relatively large values for the intercept, a. What are the likely causes of this phenomenon?

Instructor's Notes

1. Allow at least 1 dm^3 of the standard cobalt solution for every three students.

2. An analysis of variance (ANOVA) based on the results from a group of students using at least three different instruments is a useful exercise in assessing the quality of different types of instrument and the competence of different operators. Computer programs facilitate the rather complex calculations involved.

DATA SHEET FOR EXPERIMENT B.1

Least Squares Fit for a Beer-Lambert Plot (see also p. 9)

$$A = \epsilon bc + a$$

$$\epsilon = \frac{\Sigma c_i \Sigma A_i - n\Sigma c_i A_i}{(\Sigma c_i)^2 - n\Sigma c_i^2} \qquad a = \frac{\Sigma c_i \Sigma c_i A_i - \Sigma A_i \Sigma c_i^2}{(\Sigma c_i)^2 - n\Sigma c_i^2}$$

Standard Solutions

n	c_i	A_i	$c_i A_i$	c_i^2	$A_i^{calc} = \epsilon bc + a$	$\Delta = A_i^{calc} - A_i$	Δ^2
1							
2							
3							
4							
5							
6							
7							
8							
9							
10							
Σ							

Unknown Solution

n	A_i	$A_i - a$	$c_i = \dfrac{A_i - a}{\epsilon}$	$\Delta = c_i - \bar{c}_i$	Δ^2
1					
2					
3					
4					
5					
6					
7					
8					
9					
10					

Calculated Values: ϵ =   a =

$$s_1 = \left(\frac{\Sigma \Delta^2}{n-2}\right)^{\frac{1}{2}} \quad = \qquad\qquad\qquad \bar{c}_i =$$

$$s_2 = \left(\frac{\Sigma \Delta^2}{n-1}\right)^{\frac{1}{2}} \quad = \qquad\qquad \text{confidence limits} = \bar{c}_i \pm \frac{ts_2}{n^{\frac{1}{2}}}$$

B.2 The Determination of Stoichiometry of a Metal Complex by Visible Spectrometry

Object

To determine the stoichiometry of the complex formed between iron (III) and 2-hydroxybenzoic acid (salicylic acid) in aqueous solution by visible spectrometry using Job's method of continuous variation.

Introduction and Theory

Metal complexing agents are used in analytical procedures in various roles, e.g. as colour-forming reagents for quantitative analysis by visible spectrometry, as masking agents, as titrimetric and gravimetric reagents and in the solvent extraction and ion-exchange separation of metals. It is important that the stoichiometry of the reaction between a metal and a complexing agent is known for the purposes of quantitative calculations and so that analytical procedures can be properly defined. Where the complex is coloured, its stoichiometry can be established using visible spectrometry to measure the absorbances of solutions of known composition. One method that is widely applicable is Job's method of continuous variations. Separate standard solutions of the metal ion and the complexing agent are mixed in different molar proportions such that the total molarity is constant, i.e. the mole fraction of both metal ion and reagent are varied within a fixed total molarity. Absorbance readings for the series of solutions are plotted against the mole fractions of metal and reagent to give two intersecting straight lines, the point of intersection corresponding to the stoichiometry of the complex in terms of the mole ratios. It is assumed that the Beer-Lambert Law (p. 9) applies to the system.

Requirements

Iron (III) ammonium sulphate, AnalaR grade

2-Hydroxybenzoic acid (salicylic acid), GPR grade

Hydrochloric acid, 0.002M

Two 50-cm^3 burettes

Ten 100-cm^3 beakers or conical flasks

Single-beam visible spectrometer (for a filter instrument, a filter having a maximum transmittance at or near 525 nm is required)

1-cm path length glass cells (cuvettes)

Procedure

1. Prepare separate 2×10^{-3} mol dm^{-3} solutions of iron (III) ammonium sulphate and 2-hydroxybenzoic acid in 0.002M hydrochloric acid by dissolving the appropriate amounts of each in about 75 cm^3 of the hydrochloric acid in beakers. Transfer each solution quantitatively to a 250-cm^3 flask and dilute to the mark with the hydrochloric acid.

2. Prepare 9 solutions of the iron (III) complex of 2-hydroxybenzoic acid by mixing the iron (III) and acid solutions dispensed from separate burettes into beakers or conical flasks and according to the following scheme:

Volume of acid/cm³	Volume of Fe (III)/cm³	Mole fraction of acid, x
5	45	0.1
10	40	0.2
15	35	0.3
20	30	0.4
25	25	0.5
30	20	0.6
35	15	0.7
40	10	0.8
45	5	0.9

3. Measure the absorbance of each solution at 525 nm using distilled or deionized water as a reference. Measure the absorbance of the Fe(III) solution at the same wavelength.

Results

1. Calculate the molar absorptivity of Fe(III) using the Beer-Lambert relationship

$$A = \epsilon_{Fe}.b.c$$

where A is the measured absorbance of the Fe(III) solution, ϵ_{Fe} is the molar absorptivity of Fe(III), b is the path length of the cell in cm, and c is the concentration of Fe(III) in mol dm⁻³.

2. Calculate a corrected absorbance, A_c, for each of the 9 solutions prepared in 1, using the relation

$$A_c = A - [\epsilon_{Fe}(1 - x).b.c]$$

where A is the uncorrected absorbance for each of the solutions.

3. Plot a graph of A_c against x and locate the value of x where the extrapolated   straight-line portions of the graph intersect.

4. Establish the stoichiometry of the Fe(III)/2-hydroxybenzoate complex. $Fe(HB)_n$, by evaluating n in the expression

$$x = \frac{n}{1 + n}$$

Discussion

1. What is the structural formula of the Fe(III)/2-hydroxybenzoic acid complex?

2. Why is it necessary to correct the absorbance readings? How significant and how accurate are the corrections?

3. Why does the graph of mole fraction against absorbance show curvature between the linear portions?

4. Suggest alternative procedures for establishing the stoichiometry of a metal complex.

Instructor's Notes

1. This experiment could be extended by collecting data from solutions of different molarity and using a graphical method to calculate a value for the formation constant, K_f, of the metal complex.

2. Other suitable systems for study by Job's method include iron (II)/1,10-phenanthroline and cobalt (III)/nitroso-R salt (1-nitroso-2-hydroxynaphthalene-3, 6-disulphonate)

    Experimental conditions relating to these complexes are given in Colorimetric Metal Analysis, 3rd edn, E.B. Sandell, Interscience, 1959.

B.3 The Determination of Aspirin and Caffeine in a Proprietary Analgesic by Ultraviolet (UV) Spectrometry

Object

To demonstrate the use of uv spectrometry in the analysis of a two–component mixture, making use of differing absorption maxima.

Introduction and Theory

Aspirin and caffeine are common components of proprietary analgesics. Such products are subjected to strict quality control (QC) procedures to ensure consistency within specified limits. The procedures involve various instrumental techniques of which ultraviolet /visible spectrometry is of particular importance (see also Experiment B.13).

For a dilute mixture, where the components do not chemically interact in solution and where the absorbance curves of each display clearly defined and separate maxima, quantitative analysis based on the Beer-Lambert Law is feasible. A two-component mixture, such as aspirin and caffeine in a proprietary analgesic, may be analysed by making absorbance measurements at two characteristic maxima (one for each component) and solving the following pair of simultaneous equations:

$A_1 = E_1 C_1 b + E_2 C_2 b$ at wavelength λ_1, the maximum for component 1

$A_2 = E_1' C_1 b + E_2' C_2 b$ at wavelength λ_2, the maximum for component 2

where A_1 and A_2 are the total absorbances of the sample solution at λ_1 and λ_2 respectively, E_1, E_1', E_2 and E_2' are the absorptivities of aspirin and caffeine at λ_1 and λ_2, C_1 and C_2 are the w/v concentrations of aspirin and caffeine and b is the path length of the cells used.

Requirements

Aspirin (o–acetylsalicylic acid), GPR grade

Caffeine (1,3,7–trimethylxanthine), GPR grade

Sample of a proprietary analgesic, e.g. Phensic

Methanol, must have an absorbance of less than 0.1 above 250 nm

Sodium hydroxide solution, 4M

$250–cm^3$, $100–cm^3$, $50–cm^3$ and $25–cm^3$ volumetric flasks

Graduated pipettes

Scanning uv spectrometer, single–beam uv spectrometer

1–cm path length uv cells (cuvettes)

Procedure

1. Prepare a standard solution of aspirin (~375 mg) and caffeine (~95 mg) by dissolving each of them in 50 cm^3 of methanol, adding 10 drops of 4M NaOH and warming for 15 minutes on a steam–bath in a covered beaker. Cool and dilute each solution accurately to 250 cm^3 with methanol in a volumetric flask (see Instructor's Notes 1).

2. Dilute 0.25, 0.5, 1.0 and 1.5 cm^3 aliquots of the above standards to 25 cm^3 with methanol. Establish the positions of suitable absorbance maxima in the spectra of aspirin and caffeine (one for each compound) and prepare

calibration curves for each compound at each of the two wavelengths selected. Measurements should be made in 1-cm path length uv cells using methanol as a reference.

3. Crush and grind one tablet of the proprietary analgesic (if in tablet form) and dissolve in 50 cm³ of methanol, adding 10 drops of 4M NaOH and warming for 15 minutes on a steam-bath in a covered beaker.   Dilute the solution accurately to 100 cm³ with methanol.   (Excipients used in the manufacturing process may remain undissolved.)   Dilute 1.0 cm³ of the solution accurately to 50 cm³ with methanol, filter if turbid, and read the absorbance at the two wavelengths selected in 2.   Repeat the procedure for at least one more tablet.

4. Record the spectra of separate solutions of aspirin and caffeine in methanol.

Results

1. Using the calibration data, calculate values for the absorptivities based on mg cm⁻³ of aspirin and caffeine (E_1, E_1', E_2 and E_2') at the two wavelengths selected for measurements.   A linear regression analysis should be performed on each of the Beer's Law plots (see p.9).

2. Calculate the composition of the proprietary analgesic in mg of aspirin and mg of caffeine per tablet (or dose) using simultaneous equations.   A suitable computer program is a useful aid in checking these calculations.

Discussion

1. What are the chemical structures of aspirin and caffeine?

2. (a) Compare the spectra of aspirin and caffeine recorded in pure methanol and in the presence of NaOH and comment on the origins of the observed absorption bands.

    (b) How and why does NaOH affect one of these spectra?

3. Would solvent extraction be of any value in this particular determination?

4. Suggest an alternative technique for this determination.

Instructor's Notes

1. It is convenient, especially for less experienced students or large groups, to prepare standard aspirin and caffeine solutions beforehand so that procedure 1 can be omitted.

2. The molar absorptivities of aspirin and caffeine need not be calculated; concentrations are best expressed in mg cm⁻³.

3. The wavelength scales of the single-beam and scanning spectrometers may not be coincident, especially if the instruments have not been recently calibrated.   It is necessary, therefore, to check that the selected wavelengths are the same for both instruments.

4. Students should appreciate the need to check that the cells (cuvettes) used are silica (quartz), as glass cells absorb uv radiation significantly below 300 nm.

B.4 Analysis of the Composition of a Mixture of Nitroanilines by Thin-Layer Chromatography and Ultraviolet/Visible Spectrometry

Object

To identify the nitroanilines present in a mixture of isomers, and to determine the quantitative composition of the mixture using uv/visible spectrometry after separation by thin-layer chromatography.

Introduction and Theory

This experiment demonstrates how two entirely different techniques can be used in conjunction to solve a particular analytical problem.

Nitroanilines and other aromatic amines are used as intermediates in the manufacture of azo dyes and are synthesized, for example, by the nitration of acetanilide followed by hydrolysis. Such intermediates are often coloured and can be determined quantitatively by visible spectrometry through application of the Beer-Lambert law which relates absorbance and concentration (see p.9 and Experiment B.1). Where a mixture of products is formed, the separation and identification of the individual components or isomers can be accomplished readily by thin-layer chromatography on a silica gel stationary phase (see Experiment A.7). The separated isomers are removed from the thin-layer plate, dissolved in a suitable solvent and their absorptions compared with those of standards.

Requirements

2-nitroaniline (o-nitroaniline), GPR grade

3-nitroaniline (m-nitroaniline), GPR grade

4-nitroaniline (p-nitroaniline), GPR grade

Unknown mixture of nitroanilines

Industrial methylated spirits (IMS), methylbenzene (toluene), GPR grade

Chromatographic development tanks and absorbent paper

Plastic or aluminium foil backed TLC plates (without fluorescent indicator)

Thin-layer extraction device (p.61) or 10-cm^3 centrifuge cones

50-cm^3, 10-cm^3 and 5-cm^3 volumetric flasks

10 µl disposable micropipettes (microcaps) or a microsyringe

Scanning uv/visible spectrometer, single-beam uv/visible spectrometer

1-cm and 5 or 10-cm path length uv cells (cuvettes)

Procedure

1. Prepare separate stock solutions of the three nitroanilines by dissolving about 0.1 g of each (accurately weighed) in IMS and diluting to volume in 50-cm^3 volumetric flasks. Prepare a solution of the unknown mixture by dissolving 0.5 g (accurately weighed) in IMS and diluting to 100 cm^3 in a volumetric flask.

2. Line a TLC development tank with absorbent paper for about two-thirds of its height and add sufficient methyl benzene (toluene) to soak the paper and cover the bottom of the tank to a depth of about 0.5 cm. Cover the tank and allow it to stand for at least 15 minutes. Cut a 20 cm x 20 cm TLC plate in half and draw a very faint pencil line across it about 1 cm from one end. Apply a series of five 10 µl sample spots along the line

alternating the three nitroanilines with duplicates of the unknown mixture.

3. Place the plate in the tank with the sample spots at the bottom, but not
    covered by the developing solvent. Replace the cover and allow the
    solvent front to travel up the plate for about two-thirds to three-quarters
    of its length (~45 minutes). While the plate is developing, continue with
    the procedure at paragraph 5.

4. Remove the TLC plate from the tank, mark with a pencil the point to which
    the solvent front has reached and allow the plate to dry. (Gentle heating
    with a hot-air blower will help). Examine the plate to identify which
    nitroanilines are present in the unknown mixture by comparison of R_f values
    (p.10).

5. Dilute 0.5 cm³ aliquots of the stock solutions of the three nitroanilines
    prepared in 1 to volume with IMS in 100-cm³ volumetric flasks. Record
    the uv/visible spectrum of each between 325 nm and 450 nm using IMS as a
    reference. (The 2- and 3-isomers should be measured in 5-cm or 10-cm
    path length cells, the 4-isomer in a 1-cm cell.) Note the maximum in
    the absorbance curve for each isomer.

6. For those nitroanilines detected in the unknown mixture in 4, prepare three
    more standards by further dilution of those already prepared in 5. The
    concentration range of the standards should be about 0.001 mg cm⁻³ to 0.01
    mg cm⁻³.

7. Remove the spots of the nitroanilines separated from the mixture from the
    TLC plate in the following way:

    Loosen each spot from the plastic or aluminium foil backing using a spatula
    or similar instrument. Transfer the silica gel powder quantitatively to
    the thin-layer extraction device using suction (p.61) or into a 10-cm³
    centrifuge cone. If an extraction device is used, wash the silica gel
    several times with small portions of IMS, collecting the washings in the
    5-cm³ volumetric flask by applying gentle suction. If centrifuge cones
    are used, mix the silica gel with small portions of IMS, centrifuging each
    time and decanting the supernatant liquid into a 5-cm³ volumetric flask.
    Dilute the contents of each flask to volume with IMS.

8. Measure the absorbances of the standards prepared in 5 and 6 and each
    nitroaniline removed from the TLC plate at the appropriate wavelength
    maximum against IMS as a reference using a single-beam uv/visible
    spectrometer (measure the 2- and 3-isomers in 5-cm or 10-cm path length
    and the 4-isomer in a 1-cm cell.)

Results

1. Identify the nitroanilines present in the unknown mixture.

2. Prepare calibration graphs for each nitroaniline and use them to establish
    the concentrations in the duplicate sample solutions. Calculate the mean
    percentage composition of the unknown mixture.

Discussion

1. Why are different path length cells required for the uv/visible absorbance
    measurements?

2. Which steps in the procedure are likely to lead to the most significant errors in this analysis?

3. Suggest (a) an alternative means of quantification after the TLC separation and (b) an alternative technique for the entire analysis.

4. Comment on the relative merits of TLC and uv/visible spectrometry in qualitative and quantitative analysis.

Instructor's Notes

1. This is a relatively lengthy and involved experiment which is best given only to the more capable and experienced students.

2. The amount of the individual nitroanilines in unknown mixtures should be between 20 per cent and 80 per cent. A mixture of two only is recommended

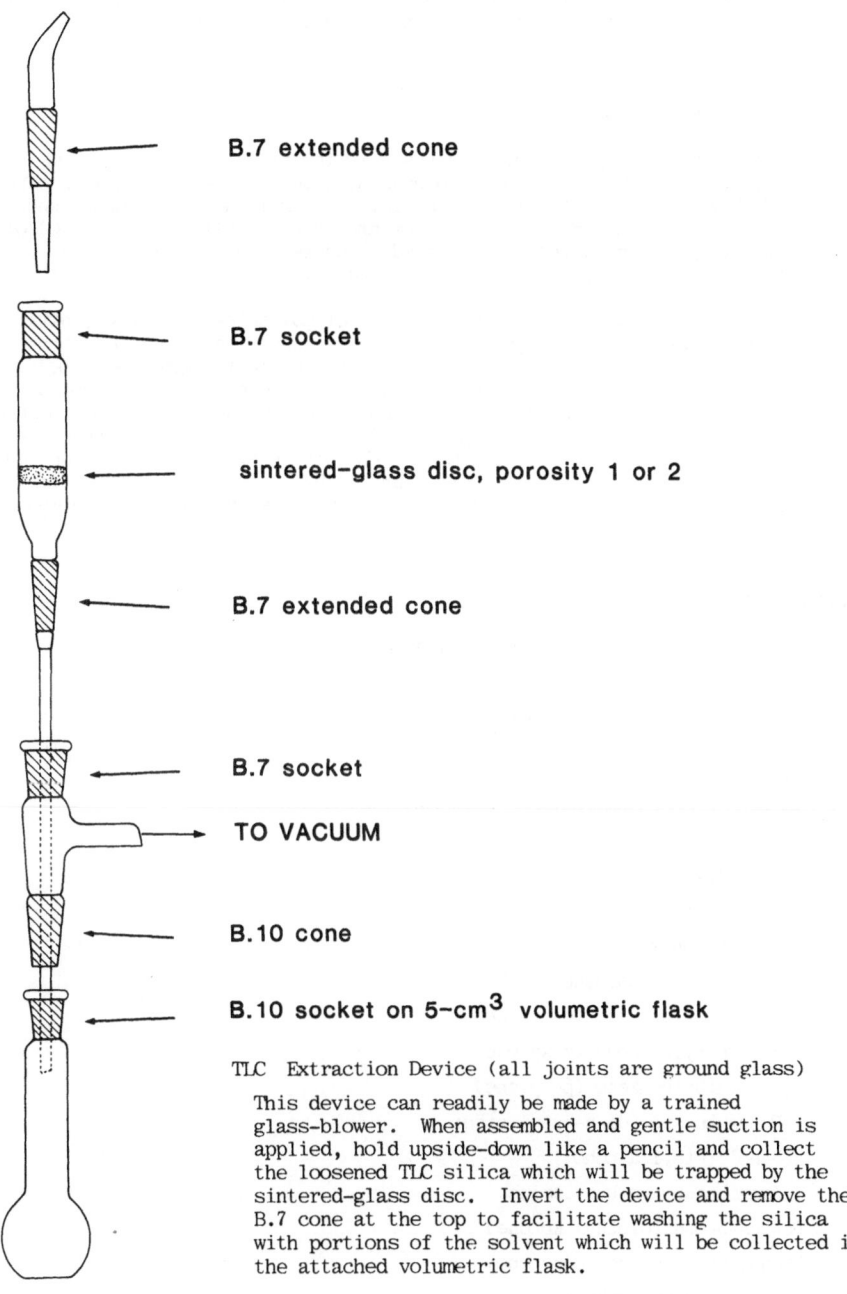

B.7 extended cone

B.7 socket

sintered-glass disc, porosity 1 or 2

B.7 extended cone

B.7 socket

TO VACUUM

B.10 cone

B.10 socket on 5-cm^3 volumetric flask

TLC Extraction Device (all joints are ground glass)

This device can readily be made by a trained glass-blower. When assembled and gentle suction is applied, hold upside-down like a pencil and collect the loosened TLC silica which will be trapped by the sintered-glass disc. Invert the device and remove the B.7 cone at the top to facilitate washing the silica with portions of the solvent which will be collected in the attached volumetric flask.

B.5 A Study of Characteristic
Infrared Absorption Frequencies

Object

To obtain and study the infrared spectra of a selection of compounds with a
range of common functional groups and molecular structures.

Introduction and Theory

Infrared spectrometry is used extensively for the identification and
structural analysis of organic compounds, often in conjunction with other
techniques such as ultraviolet/visible, nuclear magnetic resonance and mass
spectrometry.   Its main applications are in the quality control (QC) of raw
materials, intermediates and finished products, the synthesis of new compounds
and the identification of unknown substances.

Changes in the vibrational energy levels of molecules occur when infrared
radiation is passed through a sample.   The resulting spectrum of transmittance
(or absorbance) as a function of wavenumber (or wavelength) for a polyatomic
molecule is usually complex, consisting of many overlapping absorption bands
of varying intensities and widths.   For each compound, the complete spectrum
forms a unique pattern or 'fingerprint'.   The positions and intensities of
many of the bands can be correlated with particular functional groups and
other structural features.   By comparing their infrared spectra, similarities
and differences between compounds can be established and unknown compounds
partially or completely identified.

Requirements

Scanning infrared spectrometer (4000 cm^{-1} to 600 cm^{-1})

Sodium chloride or potassium bromide plates (cell windows)

0.1 mm path length liquid cell

Nujol, hexachlorobutadiene (HCBD) and/or Voltalef 3S (Kel-F), dry potassium
bromide, GPR grade

Tetrachloromethane (carbon tetrachloride)

Trichloromethane (chloroform)

Octane (n-octane)

Oct-1-ene (1-octene)

Methylcyclohexane

Methylbenzene (toluene)

1,2-Dimethylbenzene (o-xylene)

1,3-Dimethylbenzene (m-xylene)

1,4-Dimethylbenzene (p-xylene)

1-Butoxybutane (di-n-butyl ether)

Ethanol (ethyl alcohol)

1-Aminobutane (n-butylamine)

Pentan-3-one (diethyl ketone)

Phenylethanone (acetophenone)

Butanal (butyraldehyde)

Ethyl ethanoate (ethyl acetate)

Ethanoic acid (acetic acid)

Decanedioyl chloride (sebacoyl chloride)

N-methylethanamide (N-methylacetamide)

Nitrobenzene

Sodium ethanoate (sodium acetate)

4-chloroaniline (p-chloroaniline)

Procedure

1. Select at least 8 liquids from the above list and record a full-range
    (4000 cm^{-1} to 600 cm^{-1}) infrared spectrum of each as a thin film between
    sodium chloride or potassium bromide plates.   N.B. Do not use badly
    scratched, pitted or fogged plates.   Handle plates with care and by the
    edges only.

2. Select two of the compounds used in 1 and prepare separate 10% v/v
    solutions of each in tetrachloromethane.   Record a full-range infrared
    spectrum of each solution and of tetrachloromethane itself in a 0.1-mm cell.

3. Select one of the solid compounds from the above list and prepare two
    mulls, one in Nujol and one in hexachlorobutadiene or Voltalef 3S.
    Record a full-range infrared spectrum of each mull and of the mulling agent
    alone.   N.B. Thorough grinding of a few milligrams of the compound before
    mixing with a few drops of the mulling agent is essential for obtaining
    good-quality spectra.

4. Prepare a potassium bromide disc of one of the solid compounds on the above
    list (separate instruction should be available in the laboratory).   Record
    a full-range infrared spectrum of the disc and of a blank disc for
    comparison purposes.   N.B. Thorough grinding of the potassium bromide and
    the compound is essential for obtaining good-quality spectra.

Results

1. Compare the recorded spectra with available reference spectra, e.g. those
    included in the computer program "Infrared Spectrometry in Chemical
    Analysis" (Note 1) which can be run on a BBC Model B microcomputer.   Re-
    record any spectra which are not satisfactory.

2. Write down the molecular structure of each compound and identify the
    principal bands in each recorded spectrum by viewing spectra in the
    computer program or by comparison with other reference spectra.

Discussion

1. Depending on the compounds selected, comment on differences and
    similarities between the following pairs or groups of spectra.   Pay
    particular attention to the positions of specific bands:

        Octane and oct-1-ene

        Octane and methylcyclohexane

        Methylcyclohexane and methylbenzene

        The substituted benzenes

EXPERIMENT B.5

1-Butoxybutane and ethanol

Ethanol and 1-aminobutane

Ethanol and ethanoic acid

1-Aminobutane and 1-butoxybutane

Compounds containing a carbonyl group

Ethyl ethanoate, ethanoic acid and sodium ethanoate

Pentan-3-one and ethyl ethanoate

Pentan-3-one and butanal

Pentan-3-one and phenylethanone

Pentan-3-one and decanedioyl chloride

4-chloroaniline, 1-aminobutane and trichloromethane

2. Indicate the advantages/disadvantages of using tetrachloromethane as a solvent and suggest an alternative.

3. What is the purpose of recording spectra of mulls in both Nujol and hexachlorobutadiene or Voltalef 3S?

4. Why should the potassium bromide used to prepare a disc be dry and how does thorough grinding of the sample improve the quality of the spectrum?

Instructor's Notes

1. Two programs in this series are published by John Wiley and Sons Ltd, Chichester, Sussex, UK.

2. Badly scratched, pitted or fogged cell windows give unacceptably poor transmission characteristics. Regular re-polishing is recommended.

3. Students should appreciate the causes of poor-quality spectra, the need for careful handling of cells and cell windows and the need for thorough grinding of solid samples.

B.6 Quantitative Analysis by Infrared Spectrometry

Object

To determine the amount of propanone (acetone) in propan-2-ol (isopropyl alcohol) by quantitative infrared spectrometry using the carbonyl band absorption.

Introduction and Theory

Although infrared spectrometry is primarily used for qualitative analysis (See Experiment B.5), band intensities are related to the concentration and path length of the sample through the Beer-Lambert law(p.9) just as in the ultraviolet and visible regions of the spectrum.   Because of the complexity of infrared spectra and the effect of molecular interactions in solution on the positions and intensities of some bands, the selection of a suitable band upon which to base quantitative measurements is often difficult.   In addition, older instruments not having the benefits of microprocessor control and ratio-recording do not produce quantitative data of the highest precision.   However, under favourable conditions and with modern instrumentation, the analysis of two and even multicomponent mixtures is feasible.   The determination of propanone (acetone), which is a common impurity in propan-2-ol (isopropyl alcohol) as a result of facile oxidation, provides a good example:

$$(CH_3)_2CH.OH \xrightarrow{[O]} (CH_3)_2C{=}O + H_2O$$

The strongest absorption band in the infrared spectrum of propanone is the C=O stretching vibration at about 1720 cm^{-1} which is in a region where propan-2-ol has very low absorption.   A series of propanone/tetrachloromethane standards can be used to prepare a Beer's Law calibration plot based on this band and the concentration of propanone in a sample of propan-2-ol determined after suitable dilution with tetrachloromethane.

Requirements

Scanning infrared spectrometer (4000 cm^{-1} to 600 cm^{-1})

0.1 mm path length liquid cell

10-cm^3 volumetric flasks and graduated pipettes

Propanone (acetone), GPR grade

Propan-2-ol (isopropyl alcohol),GPR grade

Tetrachloromethane, GPR grade

Sample of propan-2-ol containing propanone (5 to 15% v/v)

Procedure

1. Prepare a 10% solution of the sample of propan-2-ol in tetrachloromethane and record the spectrum between 4000 cm^{-1} and 600 cm^{-1}.   If the absorbance of the carbonyl stretching band is outside the range 0.3 to 0.7 prepare another solution diluted so as to bring the carbonyl absorbance within this range.   Flush the cell several times with tetrachloromethane and handle it with care.

2. Prepare a series of five propanone standards in tetrachloromethane between 0.2% and 2.0% v/v, each in a total volume of 10 cm^3.   Record the spectrum of each standard between 1600 cm^{-1} and 1900 cm^{-1}, flushing the cell thoroughly with tetrachloromethane between each one and superimposing at

least three scans on one chart.   N.B. Do not alter the 100% control on
the spectrometer during the experiment and use the same cell throughout.

3. After calculation of the concentration of propanone in the unknown as
described in Results 2, prepare a synthetic mixture of propanone in
propan-2-ol to match the unknown exactly.   Prepare a solution of this
mixture in tetrachloromethane to match the one used in 1 above, i.e. that
which gave an absorbance between 0.3 and 0.7.   Record a spectrum of this
solution and compare it with that obtained in 1.

Results

1. For each standard and the unknown, draw a tangential baseline to the
carbonyl band and measure the difference in absorbance units between the
peak maximum and the point where a perpendicular drawn from the peak
maximum intersects the baseline (net absorbance).

2. Prepare a Beer's Law plot (p. 9) from the net absorbances and
concentrations of the standards and determine the concentration of the
unknown from the graph.   Calculate the molar absorptivity, ϵ, of the
carbonyl band in m^2 mol^{-1}.   (Density of propanone = 0.790 g cm^{-3}).

Discussion

1. Assign the principal bands in the spectra of propanone and propan-2-ol.

2. Comment on the adherence of the system to the Beer-Lambert Law.   Would
you expect similar behaviour if propan-2-ol were to be determined using
the band at 3400 cm^{-1}?

3. What alternative techniques would be feasible for this determination?

4. Comment on the molar absorptivity of the carbonyl band compared to typical
values in the ultraviolet region of the spectrum.

Instructor's Notes

1. Students should understand the relation between absorbance, transmittance
and per cent transmittance.   If per cent transmittance chart paper is
used, peak and baseline absorbances must be calculated before subtraction.

2. Badly scratched, pitted or fogged cell windows give unacceptably poor
transmission characteristics.   Regular re-polishing is recommended.
Students should appreciate the need for careful handling of cells and for
thorough flushing of cells between samples.

B.7   The Determination of Calcium in Dolomite by
Flame Emission Spectrometry (Flame Photometry)

Object

To determine the calcium content of the mineral dolomite by dissolution in
hydrochloric acid followed by flame emission spectrometry using a calibration
curve.

Introduction and Theory

Flame emission spectrometry (FES) or flame photometry is particularly suited
to the determination of alkali and alkaline earth metals in aqueous samples or
those that are readily brought into solution.   The technique is used
routinely in clinical and biochemical analysis and to a lesser extent for
agricultural samples, cements and surface waters.

If a solution containing a metal is aspirated in the form of an aerosol into a
hot flame, the solvent is evaporated from the droplets and the metal vaporized
mainly as atoms.   A proportion of these atoms will be excited by the thermal
energy of the flame and emit electromagnetic radiation characteristic of the
metal.   The intensity of the emitted radiation, measured at a suitable
wavelength, is directly proportional to the concentration of the metal in
solution and hence in the original sample.

Requirements

Calcium carbonate, AnalaR grade

Concentrated hydrochloric acid, AnalaR grade

250-cm^3 and 100-cm^3 volumetric flasks, graduated pipettes, beakers

Flame photometer with calcium filter (or flame emission spectrometer)

Sample of dolomite (from Bureau of Analysed Samples Ltd)

Procedure

1. Prepare a stock solution of calcium containing 1 mg cm^{-3} (1000 ppm) by
   suspending the appropriate amount of AnalaR calcium carbonate in about
   50 cm^3 of distilled or deionized water in a 250-cm^3 beaker and adding HCl
   dropwise until, with gentle warming, a clear solution is obtained.   Keep
   the beaker covered with a watch glass to avoid losses by spraying.
   Transfer the solution quantitatively to a 250-cm^3 volumetric flask and
   dilute to the mark.

2. Prepare a series of standard solutions (at least four) to contain between
   5 and 50 ppm of calcium by appropriate dilution of aliquots of the stock
   solution prepared in 1.   (Use small volume graduated pipettes or a
   microburette for this purpose.)

3. Weigh three separate 0.1 to 0.2 g samples of dolomite accurately into
   100-cm^3 beakers.   Add about 25 cm^3 of 1:1 HCl/water to each, cover with a
   watch glass and warm gently for about 10 minutes or until dissolution is
   essentially complete.   Filter each solution into a 250-cm^3 volumetric
   flask, and wash the filter papers several times with distilled or
   deionized water, adding the washings to the flasks.   Dilute the solutions
   to 250 cm^3.   Transfer separate 25 cm^3 aliquots of the three solutions to
   100-cm^3 volumetric flasks and dilute each to the mark.

4. Calibrate the flame photometer or spectrometer by aspirating distilled or
    deionized water into the flame and setting the reading to zero. then
    aspirating the 50-ppm standard and setting the reading to the maximum of
    the scale.

5. Aspirate each of the remaining standards and the three samples in random
    order, noting the readings and rinsing the system with water between each.
    The zero and maximum settings should be checked once or twice during this
    procedure and adjustments made if necessary.   If these readings have
    changed by more than 3 per cent, all solutions should be read again.

Results

1. Plot a calibration graph of flame photometer (spectrometer) readings
    against ppm of calcium for the series of standards measured in 4 and 5 and
    use it to determine the concentration of calcium in each of the three
    sample solutions.

2. Calculate the following:

    (a) the per cent calcium in the dolomite for each sample

    (b) the mean per cent calcium in the dolomite

    (c) the estimated standard deviation for the three results (p.17)

    (d) the confidence interval about the mean (p.17).

Discussion

1. What is dolomite?

2. Why is flame emission spectrometry particularly suitable for the
    determination of alkali and alkaline earth metals?

3. What types of interference might occur in flame emission spectrometry
    procedures? Give an example of each type.

4. Suggest alternative techniques for analysing dolomite for its major
    constituents.

B.8 The Determination of the Total Hardness and Individual Calcium and Magnesium Contents of Tap Water

Object

To determine the total calcium and magnesium content of a sample of tap water by titration with ethylenediaminetetra acetic acid (EDTA) and to determine calcium by flame emission spectrometry and hence magnesium by difference.

Introduction and Theory

This experiment demonstrates how two entirely different techniques can be used in conjunction to solve a particular analytical problem. The determination of total water hardness (Ca + Mg) is one of the routine quality control (QC) analyses carried out by Water Authorities on tap water supplies. Sometimes the proportions of calcium and magnesium as well as their sum are also of interest.

Many metals in solution can be determined by titration with a standard solution of a complexing agent of which EDTA is the most widely used. The reaction with calcium and magnesium has a 1:1 stoichiometry and is quantitative above pH 10, where EDTA is almost fully dissociated. The reagent, which can be represented as H_4Y, is normally used in the form of its disodium salt, H_2Na_2Y, as it dissolves readily in water, the undissociated acid being much less soluble. Metallochromic indicators such as eriochrome black-T, form coloured complexes with metal ions but change colour when the metals are fully complexed with EDTA as at the end-point in a titration. As these indicators are also senstive to pH changes, solutions to be titrated must be well-buffered.

When both calcium and magnesium are present in the sample, the EDTA titre gives the sum of the two, i.e. the 'total hardness' of the water. The calcium content can be determined independently by flame emission spectrometry (FES), or flame photometry, a technique that is particularly suited to the determination of alkali and alkaline earth metals. If a solution containing a metal is aspirated in the form of an aerosol into a flame, the solvent is evaporated from the droplets and the metal vaporized mainly as atoms. A proportion of metal atoms will be excited by the thermal energy of the flame and emit electromagnetic radiation characteristic of the metal. The intensity of the emitted radiation, measured at a suitable wavelength, is directly proportional to the concentration of the metal in the solution and hence in the original sample.

Requirements

Ethylenediaminetetraacetic acid, disodium salt, AnalaR grade

Eriochrome black-T indicator solution, 1% w/v in 3:1 triethanolamine/ethanol

Calcium carbonate, AnalaR grade

Ammonia-ammonium chloride buffer solution pH 10

(142 cm^3 of concentrated aqueous ammonia, sp.gr.0.880, and 17.5 g of AnalaR grade ammonium chloride diluted to 250 cm^3 with distilled or deionized water)

Potassium cyanide, AnalaR grade —CAUTION, VERY TOXIC, HANDLE WITH CARE

Hydroxyammonium chloride, AnalaR grade

Burette, graduated pipettes, 250-cm^3 and 100-cm^3 or 50-cm^3 volumetric flasks

Flame photometer with calcium filter or flame emission spectrometer

Procedure

1. Prepare a 0.01 mol dm^{-3} solution of EDTA by dissolving the appropriate amount in about 100 cm^3 of distilled or deionized water with warming. Cool, transfer the solution quantitatively to a 250-cm^3 volumetric flask and dilute to the mark.

2. Transfer duplicate 50 cm^3*samples of tap water to 250-cm^3 conical flasks and acidify each with 2 or 3 cm^3 of dilute HCl. Boil for one minute, cool and neutralize with dilute aqueous NaOH. Add 10 cm^3 of the ammonia-ammonium chloride buffer solution to each and a few crystals each of potassium cyanide and hydroxyammonium chloride.

3. Add 3 to 5 drops of eriochrome black-T indicator solution and titrate each sample with 0.01 mol dm^{-3} EDTA to a red→blue end-point. Repeat if necessary on further 50 cm^3* samples of tap water until titres agree to within 0.5 per cent.

4. Prepare a stock solution of calcium containing 1 mg cm^{-3} (1000 ppm) by suspending the appropriate amount of AnalaR calcium carbonate in about 50 cm^3 of distilled or deionized water and adding dilute HCl dropwise until,with gentle warming, a clear solution is obtained. Use a 250-cm^3 beaker covered with a watch glass to avoid losses by spraying. Transfer the solution quantitatively to a 250-cm^3 volumetric flask and dilute to the mark.

5. Prepare a series of standard solutions (at least four) to contain between 5 and 50 ppm of calcium by appropriate dilution of aliquots of the stock solution prepared in 4. (Use small volume graduated pipettes for this purpose.)

6. Calibrate the flame photometer or flame emission spectrometer by aspirating distilled or deionized water into the flame and setting the reading to zero, then aspirating the 50 ppm standard and setting the reading to the maximum of the scale. Aspirate the remaining standards in turn, noting the readings and rinsing the system by aspirating water for about 10 seconds between each one.

7. Aspirate undiluted tap water and note the reading. Check that the instrument response is stable by aspirating one or more standards again. If necessary repeat the procedure until all readings are repeatable to within 3 per cent.

Results

1. From the EDTA titration results, calculate the total hardness of tap water in terms of mg dm^{-3} of calcium carbonate.

2. Plot a calibration graph of instrument readings against ppm of calcium for the series of standards measured in 6, and use it to determine the concentration of calcium in the tap water in terms of mg dm^{-3} of calcium carbonate.

3. Calculate the magnesium content of the tap water by difference using the results for total hardness and calcium.

* The sample size can be varied to suit the degree of hardness.

Discussion

1. In what chemical forms do calcium and magnesium occur in tap water?

2. Why are the following steps necessary before titrating the samples with EDTA?

    (a) acidifying, boiling and neutralizing

    (b) buffering the solution to pH 10

    (c) adding potassium cyanide and hydroxyammonium chloride.

3. Why is flame photometry (or FES) particularly suitable for the determination of alkali and alkaline-earth metals?

4. Would it be feasible to determine calcium and magnesium separately by titration with EDTA?

Instructor's Notes

1. Some tap waters may have a very low magnesium level. As no distinct end-point is observed for the titration of calcium using eriochrome black-T, a small amount of the magnesium-EDTA complex should be added to the buffer solution or the tap water sample before titration. This results in a much sharper end-point colour change. (See A Textbook of Quantitative Inorganic Analysis, 4th edn., A.I. Vogel, Longman, 1978).

2. Sutdents should be encouraged to be critical of dirty volumetric glassware and to use a white background to observe the end-points of titrations more clearly.

3. Some students may require guidance in calculating the weight of calcium carbonate to be used in preparing the 1000 pm calcium stock solution and in converting results to mg dm^{-3} of calcium carbonate.

B.9 The Determination of Nickel in Steel
by Atomic Absorption Spectrometry

Object

To determine the nickel content of a steel sample by atomic absorption
spectrometry (AAS) using a calibration curve, and by the method of standard
addition.

Introduction and Theory

Atomic absorption spectrometry (AAS) is used widely for the quantitative
determination of metals as minor or trace constituents of samples as varied as
alloys, rocks and soils, foodstuffs and drinks, surface waters, biological
fluids and reagent chemicals.

If a solution containing a metal is aspirated in the form of an aerosol into a
hot flame, the solvent is evaporated from the droplets and the metal vaporized
mainly as atoms.   Alternatively, an atomic vapour can be produced by rapid
electrothermal heating of a graphite rod or tube on which a drop of the sample
has been placed.   A beam of electromagnetic radiation characteristic of a
particular element can be passed through the atomic vapour and monitored by a
photomultiplier detector.   If the sample contains that particular element,
its atoms will selectively absorb some of the radiation thereby attenuating
the beam and causing the detector signal to fall.   This absorbance is
proportional to the concentration of that element in the vapour and hence in
the original sample.

Requirements

Ammonium nickel sulphate, AnalaR grade

Concentrated nitric acid, AnalaR grade

25-cm^3 pipette, graduated pipettes, 250-cm^3 and 100-cm^3 volumetric flasks,
beakers

Steel sample (1% to 5% nickel) e.g. from Bureau of Analysed Samples Ltd

Atomic absorption spectrometer (fitted with an air/acetylene burner)

Nickel hollow-cathode lamp

Procedure

1. Prepare a stock solution of nickel containing 1 mg/cm^3 (1000 ppm) by
    dissolving the appropriate amount of ammonium nickel sulphate in about
    100 cm^3 of distilled or deionized water in a beaker with warming.   Cool
    and transfer the solution quantitatively to a 250-cm^3 volumetric flask and
    dilute to the mark with distilled or deionized water.

2. Weigh accurately two 0.2 g samples of the nickel-containing steel into
    50-cm^3 beakers.   Add 10 cm^3 of 1:1 HNO$_3$/water to each beaker, cover with
    a watch glass and warm until the samples have dissolved.   Cool, transfer
    each solution quantitatively to a 250-cm^3 volumetric flask and dilute to
    the mark with distilled or deionized water.   Dilute 25 cm^3 of each sample
    solution to 100 cm^3 in volumetric flasks and label them C1 and C2.

3. Pipette two further 25 cm^3 aliquots of each sample solution into 100-cm^3
    volumetric flasks, add 0.5 cm^3 and 1.0 cm^3 respectively of the nickel
    stock solution prepared in 1 to each flask and dilute to the mark with
    distilled or deionized water.   Label these solutions SA1 and SA2.

4. Prepare a series of standard solutions (at least four) to contain between 1 and 15 ppm of nickel by appropriate dilution of aliquots of the nickel solution prepared in 1.   (Use small volume graduated pipettes for this purpose.)

5. Adjust the atomic absorption spectrometer according to the manufacturer's instructions for the determination of nickel, selecting the hollow-cathode lamp emission line at 232 nm.   This should normally be done under supervision or by an instructor.

6. Aspirate distilled or deionized water into the flame and zero the absorbance scale.   Aspirate each standard solution and the two sample solutions C1 and C2 in turn, rinsing with distilled or deionized water between each one, and record the absorbance reading in each case.

7. Check the zero absorbance setting by aspirating water again then aspirate each of the sample solutions C1 and C2 and each of the standard addition solutions SA1 and SA2.   Record the absorbance readings in each case.

Results

1. Plot a calibration graph of absorbance against ppm of nickel for the series of standards prepared in 4, and use it to determine the concentration of nickel in solutions C1 and C2 from the absorbance readings measured in 6. Calculate the per cent nickel in the steel sample.

2. Plot a standard addition graph (p. 5) using the absorbance values for solutions C1, C2, SA1 and SA2 measured in 7.   Calculate the per cent nickel in the steel sample.

Discussion

1. Compare the results obtained using the calibration graph and by the standard addition method. What advantage, in general, is to be expected from using a standard addition procedure?

2. What changes in the procedure would be necessary to determine nickel at a lower or higher level in a steel sample?

3. Which instrumental parameters require optimization in any atomic absorption analysis?

4. Can the matrix element, i.e. iron, be considered as completely non-interfering in this procedure?

Instructor's Notes

1. Some steel samples may leave a slight residue of silica on dissolution in nitric acid.   This can normally be ignored or removed by filtration as it has no significant effect on the determination of nickel.

2. Some students may require guidance in calculating the weight of ammonium nickel sulphate to be used in preparing the 1000 ppm nickel stock solution and in diluting solutions.

B.10   The Determination of Iron in Canned Drinks
by Atomic Absorption Spectrometry

Object

To determine trace levels of iron in canned beer and cola drinks
by atomic absorption spectrometry (AAS) using a calibration curve.

Introduction and Theory

Atomic absorption spectrometry (AAS) is used widely for the quantitative
determination of metals as minor or trace constituents of samples as
varied as alloys, rocks and soils, foodstuffs and drinks, surface waters,
biological fluids and reagent chemicals.

If a solution containing a metal is aspirated in the form of an
aerosol into a hot flame, the solvent is evaporated from the droplets and the
metal vaporized mainly as atoms.
Alternatively, an atomic vapour can be produced by rapid electrothermal
heating of a graphite rod or tube on which a drop of the sample has been
placed.   A beam of electromagnetic radiation characteristic of a particular
element can be passed through the atomic vapour and monitored by a photomultiplier detector.   If the sample contains that particular element, its
atoms will selectively absorb some of the radiation, thereby attenuating the
beam and causing the detector signal to fall.   This absorbance is
proportional to the concentration of that element in the vapour and hence in
the original sample.

Requirements

High-purity iron granules (Bureau of Analysed Samples Ltd)

Concentrated hydrochloric acid, AnalaR grade

Measuring cylinder, 100-cm^3

Pipettes and volumetric flasks

Samples of a canned beer and a cola drink

Atomic absorption spectrometer (fitted with air/acetylene burner)

Iron hollow-cathode lamp

Procedure

1. Prepare a stock solution of iron containing 1 mg/cm^3 (1000 ppm) by the
    following procedure:

    Weigh exactly 0.5 g of high-purity iron granules into a 100-cm^3 beaker
    and dissolve them in the minimum quantity of 1:1 HCl/water, covering the
    beaker with a watch glass and heating it to aid dissolution.   Cool,
    transfer the solution quantitatively to a 500-cm^3 volumetric flask and
    dilute to the mark with distilled or deionized water.

2. Dilute an aliquot of the stock solution to provide a 2 ppm iron standard,
    making the diluted solution 0.1M with respect to HCl.   Prepare a series
    of at least four more standards covering the range 0.05 to 1.0 ppm of iron
    and 0.1M in HCl by appropriate dilution of the 2 ppm standard.

3. To 20 cm^3 of the beer in a 100-cm^3 measuring cylinder, add 20 cm^3 of distilled or deionized water and pour the contents of the cylinder into a 100-cm^3 beaker.
   To 20 cm^3 of the cola drink in a 100-cm^3 measuring cylinder, add 80 cm^3 of distilled or deionized water and pour the contents into a 150-cm^3 beaker. Stir the contents of both beakers well to assist the expulsion of carbon dioxide.

4. Adjust the atomic absorption spectrometer according to the manufacturer's instructions for the determination of iron, selecting the hollow-cathode lamp emission line at 248.3 nm. This should normally be done under supervision or by an instructor.

5. Aspirate distilled or deionized water into the flame and zero the absorbance scale. Aspirate each standard solution and the sample solutions in turn, rinsing with distilled or deionized water between each one, and recording the absorbance reading in each case. Check the zero-absorbance reading by aspirating water again, then obtain duplicate readings from the standards and sample solutions.

Results

1. Plot a calibration graph of absorbance against ppm of iron for the series of standards prepared in 2, and use it to determine the concentration of iron in each of the sample solutions.

2. Calculate the mean ppm of iron in the undiluted canned drinks.

Discussion

1. Why is it necessary to prepare iron standards in dilute acid?

2. Why does the calibration graph curve towards the concentration axis?

3. Why is it necessary to expel carbon dioxide from the samples?

4. Comment on any difference in the level of iron found in the beer and cola samples.

Instructor's Notes

1. The experiment could be shortened by using a ready-prepared 1000 ppm iron stock solution. Such solutions are commercially available.

2. A statistical analysis of class or group results is a useful exercise.

B.11 Gas Chromatographic Separation of Alkanes

Object

To separate and determine the composition of a mixture of n-alkanes and to identify unknown members of a homologous series by gas-liquid chromatography.

Introduction and Theory

Gas chromatography (GC) is used for the separation and quantitative analysis of mixtures where the components are sufficiently volatile and thermally stable to be passed through a chromatographic column in the vapour state. This normally requires elevated temperatures of 100 to 400°C. It is used in the analysis of petrochemicals and many products based on them, solvents, volatile natural products, pesticide and herbicide residues, and paints and polymers after pyrolysis. The components of a mixture are carried through the column by an inert carrier gas (usually nitrogen) and are generally eluted in order of increasing boiling points, although differing affinities for the stationary phase may affect the order of elution. The eluted compounds are detected by monitoring a physical property of the gas stream leaving the column, such as degree of induced ionization, thermal conductivity or emission of characteristic electromagnetic radiation. Eluted compounds are characterized by their retention times, t_R, and quantitative analysis is accomplished by comparing the areas of analyte peaks with those of standards.

The separation and quantitative analysis of a mixture of n-alkanes and the identification of unknown alkanes illustrates some of the capabilites of GC.

Requirements

n-octane, n-decane, n-dodecane, GPR grade

Mixture of unknown n-alkanes

Diethylether, GPR grade

1-µl capacity microsyringe

Gas chromatograph (preferably fitted with a flame ionization detector and a temperature programmer)

Apiezon-L packed column, 2 m, 10% on Chromosorb WHP or an equivalent

Procedure

1. Prepare a mixture of known composition by weight of the n-alkanes using about 20 mg of octane, 30 mg of decane and 40 mg of dodecane. Dilute the mixture about 20:1 with diethyl ether.

2. Set the chromatograph oven to between 120° and 130°C and the carrier gas flow rate (N_2) to about 25 cm^3 min^{-1}. When the oven temperatures has stabilized, inject samples of the mixture prepared in 1 until the appropriate detector attenuation setting for the <u>octane</u> peak to be almost full-scale on the chart recorder is found (the diethylether peak will be off-scale at this setting). Obtain one chromatogram at a faster chart speed to facilitate the measurement of peak areas.

3. Inject samples of the mixture of unknown alkanes until the appropriate detector attenuation setting for both peaks to be on scale is achieved and at the original chart speed.

4. If the chromatograph is fitted with a temperature programmer, set the starting temperature to 110°C and allow the oven temperature to stabilize. Set the rate of temperature increase to 8 to 10°C/min and the final temperature to 170°C and inject a sample of the mixed n-alkanes as in 2.

Results

1. Measure the retention time, t_R or volume V_R (t_R and V_R are directly proportional) for each n-alkane and the unknowns as the distance from the start of the solvent (ether) peak to the apex of each alkane peak.

2. Plot a graph of $\log_{10} t_R$ against the carbon number for the mixed n-alkanes and from it deduce the identity of the unknown alkanes.

3. Determine the column efficiencies(N) for each of the three n-alkanes in the mixture using the formula

$$N = 5.54 \, (t_R / W_{h/2})^2$$

where $W_{h/2}$ is the width at half-height measured in the same units as t_R (see p.11).

Calculate the corresponding plate heights (H) from the formula $H = L/N$, where L is the column length in mm.

4. Measure the peak areas of the three n-alkanes by 'triangulation' and calculate the percentage composition of the mixture by 'internal normalization', i.e. express each area as a percentage of the sum of all three areas (see p. 6).   Repeat the calculation using an alternative method of peak area measurement, e.g. cutting and weighing, height x width at half-height, computing integrator.   Compare both sets of results with the original composition by weight.

Discussion

1. What primarily determines the elution order in gas chromatography?

2. Why is an Apiezon-L column used to separate n-alkanes?

3. Comment on the methods of peak area measurement selected.   Which method is to be preferred and why?

4. What are the advantages of temperature programming compared to isothermal operation?

Instructor's Notes

1. Students may require guidance in the technique of filling a microsyringe so as to avoid injecting a significant quantity of air with the sample.

2. Confusion may be caused by overlapping chromatograms which arise when a sample is injected before the previous one has been completely eluted.

3. Procedure 4 is optional as many older chromatographs do not have this facility.

B.12 The Characterization of Peppermint Oils by Gas Chromatography and the Identification of Oils used in Consumer Products

Object

To compare the chromatograms of peppermint oils under different operating conditions and to identify the oils used in some consumer products.

Introduction and Theory

Gas chromatography (GC) is used for the separation and quantitative analysis of mixtures where the components are sufficiently volatile and thermally stable to be passed through a chromatographic column in the vapour state. This normally requires elevated temperatures of 100 to 400°C. It is used in the analysis of petrochemicals and many products based on them, solvents, pesticide and herbicide residues, and paints and polymers after pyrolysis. It is of particular value in the characterization of complex mixtures such as volatile natural products. Where such products are used commercially, apart from quality control (QC), it may be necessary to distinguish between similar materials of differing quality and value, to check for adulteration of one material with a cheaper and inferior alternative, or to compare a natural product with a synthetic one.

The components of a mixture are carried through the column by an inert carrier gas (usually nitrogen) and are generally eluted in order of increasing boiling points, although differing affinities for the stationary phase may affect the order of elution. The eluted compounds are detected by monitoring a physical property of the gas stream such as degree of induced ionization, thermal conductivity or emission of characteristic electromagnetic radiation. Eluted compounds are characterized by their retention times, t_R, and quantitative analysis is accomplished by comparing the areas of analyte peaks with those of standards.

Requirements

Ethyl ethanoate (ethyl acetate), AnalaR grade

American (<u>Mentha piperita</u>) and Chinese (<u>Mentha arvensis</u>) peppermint oils, 99% pure or better (see Instructor's Note 1)

Separate solutions of each peppermint oil in ethyl ethanoate, about 0.1% v/v

Consumer products, e.g. Polo, Trebor, extra strong mints, mint tea, mint toothpastes

1-μl capacity microsyringe

Gas chromatograph (fitted with a flame ionization detector and preferably with a temperature programmer)

Carbowax 20M packed column, 2 m, 10% loading on Chromosorb WHP or an equivalent

Bench centrifuge

Procedure

1. Set the chromatograph oven at 180°C and the carrier gas flow rate (nitrogen) to about 30 cm^3 per minute. When the oven temperature has stabilized, obtain chromatograms of one of the oils by injecting 0.5 to 1 ul samples until the appropriate detector attenuation (sensitivity) setting for the largest peak to be almost full scale on the chart recorder is found.

2. Inject a 0.5 μl sample of the other oil under the same conditions, then
    obtain chromatograms of both oils at two other isothermal temperatures
    between 140°C and 180°C. Allow the oven temperature to stabilize for
    several minutes after altering the setting and before injecting a sample.

3. If the chromatograph is fitted with a temperature programmer, set the
    starting temperature to 130°C, the final temperature to 180°C and the
    rate of temperature increase to between 3°C and 5°C per minute. Allow
    the oven temperature to stabilize at 130°C and inject a sample of one of
    the oils, simultaneously starting the temperature programme. On complete
    elution of the sample, cool the oven back to 130°C and obtain a chromatogram
    of the other oil in the same way.

4. Whilst the samples are being eluted in 1 to 3 above, prepare extracts of
    several of the consumer products. For the mints, crush and grind one sweet
    in a pestle and mortar. Transfer the powder to a 30-cm³ beaker and add
    sufficient AnalaR ethyl ethanoate (ethyl acetate) to produce a thin creamy
    paste (1 or 2 cm³). After thorough mixing, pour the paste into a small
    tube and centrifuge for one minute. Decant the supernatant liquid into a
    sample tube. Other products such as mint tea leaves can be treated
    similarly but mint toothpastes should be suspended in water and extracted
    with a small volume of ethyl ethanoate. This can be reduced to 1 cm³ or
    less by evaporation in a stream of nitrogen with gentle (< 50°C) warming.

5. Set the oven to one of the isothermal temperatures used in 1 and 2 or to
    130°C if a temperature programmer is available.
    Obtain chromatograms of each extract prepared in 4 and of the solutions of
    the oils by injecting 1 μl samples, setting the detector attenuation so
    that the largest peak is almost full-scale deflection (a 10- to 100-fold
    increase in sensitivity compared to the setting used for the solutions of
    the pure oils will be required).

Results

1. Compare the chromatograms of the pure oils at each isothermal
    temperature and under temperature programming conditions. Note the
    retention times of at least 10 peaks in each chromatogram and their
    relative sizes.

2. Identify the type of peppermint oil that has been used in each of the
    consumer products by comparisons of the respective chromatograms with
    those of the ethyl ethanoate solutions of the pure oils.

Discussion

1. List the types of compound of which peppermint oils are composed and
    name the two major constituents.

2. Why is GC particularly useful for the characterization of peppermint oils
    and why is Carbowax 20M a more suitable stationary phase than Apiezon-L?

3. What are the advantages/disadvantages of temperature programming compared
    to isothermal elution?

4. Comment on the types of oil found in the consumer products analysed.

Instructor's Notes

1. Samples of pure peppermint oils can be obtained from the larger
    chemist's shops (drug stores) or from Schuco Scientific Ltd., Halliwick
    Court Place, Woodhouse Road, London N12, who are agents for the German
    suppliers Carl Roth, GmBH.

2. Complete elution of the oils will take up to 20 minutes, depending on the oven temperature and carrier gas flow rate. Students often inject a second sample before the previous one has completely eluted.

3. If necessary, sensitivity can be improved by reducing the ethyl ethanoate extract volume to ~ 0.1 cm^3 as described in procedure 4.

4. If a capillary column with a moderately polar stationary phase is available, a comparison of its performance with that of the packed column provides an interesting extension to this experiment.

B.13 Quantitative Analysis by High Performance Liquid Chromatography

Object

To determine aspirin and caffeine quantitatively in a proprietary analgesic by HPLC separation on a bonded-phase column and using a UV absorbance monitor.

Introduction and Theory

Pharmaceutical products are subjected to strict quality control (QC) procedures to ensure consistency of the formulation within specified limits. Various instrumental techniques are used of which high performance liquid chromatography (HPLC) and ultraviolet spectrometry are of particular importance. Aspirin and caffeine are common components of proprietary analgesics and may be separated and determined quantitatively by HPLC (see also Experiment B.3) using paracetamol (4-acetamidophenol) as an internal standard.

HPLC is used for the separation and quantitative analysis of a wide variety of mixtures, expecially those where the components are insufficiently volatile and/or thermally stable to be separated by gas chromatography (GC). It is used extensively in the analysis of pharmaceutical products, foodstuffs and beverages, agrochemicals, polymers and plastics and for monitoring drugs and their metabolites in body fluids. The components of a mixture are carried through the column by a mobile liquid phase pumped under pressure. The order of elution is determined by the chemical nature of the components, the mobile phase and the stationary phase. Stationary phases are silica or chemically modified silica (bonded phases) of very small particle size (3 μm to 10 μm). The eluted components are detected by monitoring the UV absorbance or fluorescence, the current generated by a redox reaction (amperometry) or the refractive index of the mobile phase. Eluted components are characterized by their retention times, t_R, or their capacity factors, k', and quantitative analysis is accomplished by comparing the areas of analyte peaks (or sometimes their heights) with those of standards.

Requirements

Aspirin (o-acetylsalicylic acid), GPR grade

Caffeine (1,3,7-trimethylxanthine), GPR grade

Paracetamol (4-acetamidophenol), GPR grade

Sample of a proprietary analgesic, e.g. Phensic

Phosphate buffer solution, pH 7 (0.895 g of Analar $Na_2HPO_4.12 H_2O$ in 1 dm³ of distilled or deionized water, adjusted to pH7 with H_3PO_4)

Methanol, must have an absorbance of less than 0.1 above 250 mm

Tartrazine (C.I acid yellow), 0.005% w/v aqueous solution

50-cm³ and 25-cm³ volumetric flasks, 1.0-cm³ pipette

Liquid chromatograph with UV absorbance monitor

Spherisorb octadecyl (ODS or C18) column, 10 cm/5 μm particle size or 20 cm/ 10 μm particle size (see Instructor's Notes 1)

Procedure

1. Prepare the mobile phase (solvent) by mixing 200 cm³ of methanol with 50 cm³ of the phosphate buffer. Degas about 130 cm³ of this solvent by vigorous refluxing for 15 minutes and cooling or by the passage of helium

gas for about 1 minute.   Set the UV monitor at 254nm, select a flow-rate
of 1 to 1.5cm^3 min-1 and pump the solvent through the column until a steady
baseline is obtained on the chart record.

2. Weigh accurately about 0.7 g of aspirin (o-acetylsalicylic acid) and about
    0.1 g of caffeine into a 50-cm^3 beaker and add about 5 cm^3 of chloroform.
    Mix well for several minutes then add about 10 cm^3 of methanol and stir
    until dissolution is complete.   Transfer the solution quantitatively to a
    50 cm^3 flask and dilute to the mark with methanol.

3. Weigh about 0.045g of paracetamol (4-acetamidophenol) into a 50 cm^3 beaker
    and dissolve it in a little methanol.   Transfer the solution
    quantitatively to a 50-cm^3 flask and dilute to the mark with methanol.

4. Pipette 1.0 cm^3 of the aspirin/caffeine solution prepared in 2 and 1.0 cm^3
    of the paracetamol solution prepared in 3 into a 50-cm^3 flask and dilute to
    the mark with some of the remaining mobile phase prepared in 1.

5. Crush and grind one tablet of the proprietary analgesic, add about 5 cm^3
    of chloroform and mix well for several minutes.   Add about 5 cm^3 of
    methanol and transfer the solution and any residual excipients
    quantitatively to a 25-cm^3 flask.   Dilute to the mark with methanol and
    mix thoroughly.   Allow the excipients to settle out then transfer 1.0 cm^3
    of the supernatant solution to a 50-cm^3 flask.   Add 1.0 cm^3 of the
    paracetamol solution prepared in 3 and dilute to 50-cm^3 with some of the
    remaining mobile phase prepared in 1.

6. Obtain duplicate chromatograms of the standard solution of aspirin/caffeine
    /paracetamol prepared in 4 and duplicates of the tablet solution from 10µl
    injections via the sample valve. Obtain a chromatogram of tartrazine.

Results

1. Calculate the efficiencies(plate numbers, N) of the aspirin, paracetamol
    and caffeine peaks in the chromatograms of the standard using the formula

$$N = 5.54 \ (t_R/W_{h/2})^2$$

where t_R is the retention time measured from the point of injection to the
apex of each peak, and $W_{h/2}$ is the width at half-height for each peak
measured in the same units as t_R (see p. 11).   Calculate the corresponding
plate heights (H) from the formula H = L/N, where L is the column length
in mm.

2. Calculate the capacity factor, k', for aspirin, paracetamol and caffeine
    using the formula

$$k' = \frac{t_R - t_0}{t_0}$$

where t_0 is the retention time for an unretained substance, e.g. tartrazine

3. Calculate the amounts of aspirin and caffeine in the tablet in milligrams
    from peak area measurements using paracetamol as an internal standard
    (p. 6).   Repeat the calculation using the aspirin and caffeine peaks
    only.

Discussion

1. What is the purpose of using an internal standard? Comment on the accuracy of the results calculated with and without the internal standard.

2. What would be the effect of monitoring the UV absorbance at wavelengths other than 254 nm?

3. Rationalize the elution order from a consideration of the chemical nature of aspirin, caffeine, paracetamol and the stationary and mobile phases.

4. How would altering (a) the proportions of methanol and buffer and (b) the buffer pH be likely to affect the elution times and order of elution?

Instructor's Notes

1. The selectivity of ODS (C18) column packings from different manufacturers may vary appreciably. Alternatives to Spherisorb may therefore not separate the three compounds satisfactorily under the conditions specified.

2. Aspirin and aspirin solutions decompose slowly to release salicylic acid for which an additional chromatographic peak may be observed.

3. An investigation of the effect of pH and buffer/methanol proportions could form a useful extension to this experiment.

4. Statistical analysis of class or group results and/or comparison of results with those from Experiment B.3 would be useful exercises.

B.14 The Identification of Raw Fish Samples by Electrophoresis

Object

To identify the species of fish in raw samples by comparison of the protein patterns produced by vertical slab electrophoresis.

Introduction and Theory

Electrophoresis is a technique used widely in clinical and biochemical analysis primarily for the separation and identification of proteins, enzymes, viruses and nucleic acids. It depends upon differential rates of migration of charged species in an electrolyte solution under the influence of an applied potential gradient. The solution is retained in a solid medium, generally a polyacrylamide or agarose gel, through which the sample components migrate during application of a dc potential and which minimizes diffusional spreading of the component bands. Their positions at the end of this period are recorded by a staining process that produces coloured lines or bands in the gel, in the case of proteins the patterns being characteristic of the organism from which the sample originated. Identifications are made by comparisons with standards run at the same time.

Requirements

Glycine, GPR grade

Tris(hydroxymethyl) aminomethane (TRIZMA base)

N,N,N',N'-tetramethylethylenediamine (TEMED) ⎫ These two reagents should

Ammonium persulphate, GPR grade ⎭ have been recently purchased

Acrylamide, GPR grade

N,N'-methylenebisacrylamide, GPR grade

Glycerol, GPR grade

Bromophenol blue, 1% w/v aqueous solution

Gel stain, e.g. naphthalene black, 1% w/v in 7% aqueous acetic acid

De-staining solution, 7% aqueous acetic acid

Raw fish, samples and standards

Gel-forming template and masking tape

Vertical slab electrophoresis unit

Gel staining and de-staining dishes

Top-drive homogenizer

Bench-top centrifuge

Microsyringe or micropipette, 50µl capacity

Procedure

1. Prepare a stock solution of glycine (28.8 g) and TRIZMA base (6 g) in 1 dm^3 of distilled or deionized water adjusting the pH to 8.6 with 1M HCl. Dilute the solution to 11 dm^3 with water. (This provides sufficient buffer solution for about 5 electrophoresis runs.)

2. Prepare a gel slab in the following manner:
   Assemble the gel-forming template by taping together the two sheets of

glass, separating them with a spacer at each side and one long one across
the bottom.   Dissolve 0.25 g of N,N'-methylenebisacrylamide, 5.75 g of
acrylamide and 0.05 g of ammonium persulphate in 100 cm^3 of the pH 8.6
buffer solution and degas the solution using a vacuum pump.   N.B. This
solution must not become aerated.   Add 200 to 400 μl of TEMED, and <u>quickly</u>
pipette the gel solution into the template ensuring that no air bubbles are
trapped.   Fit the sample-well spacer/former into the top and allow the
gels to polymerize (about 10 to 15 minutes).   If the template leaks
during this period, top it up at the edges, ensuring that no air bubbles
are introduced.

3.   For each unknown fish sample and standard, weigh approximately 1 g and
     fragment it in a pestle and mortar.   Add an equal volume of distilled or
     deionized water and homogenize for 2 minutes.   Decant and discard the
     excess liquid and centrifuge the residue at full speed for 20 minutes.
     Decant the supernatant liquid and store it at 0 to 4°C.

4.   Remove the sample-well spacer/former and the bottom spacer from the
     template and transfer it to the electrophoresis unit securing it with the
     clips provided.   Fill the lower and upper reservoirs (in that order) with
     pH 8.6 buffer solution, ensuring that no air bubbles are trapped, and
     pre-run the gel at 30 mA for 20 minutes.

5.   Add 1 drop of bromophenol blue solution and several drops of glycerol to
     each standard and sample and shake vigorously.   Using a microsyringe or
     micropipette, layer 10 to 50 μl samples into the sample wells.   Run the
     gel for 45 to 60 minutes at approximately 35 mA or until the bromophenol
     blue is about three-quarters of the way down the gel slab.

6.   Remove the template and detach one of the glass plates sandwiching the gel
     slab.   Lay the other plate with the gel uppermost in a dish and cover the
     gel with the staining solution.   Leave for 15 minutes then transfer the
     plate and gel to a second dish containing 7% aqueous acetic acid.   Replace
     the acetic acid at least twice more then leave until the protein bands are
     visible and the background is essentially free of stain (this generally
     takes 24 to 48 hours).

Results

1.   Identify the unknown fish samples by comparison of their electrophoretograms
     with those of the standards.

2.   Make a diagrammatic sketch of the protein band patterns as permanent.
     record, or photograph the gel slab.

Discussion

1.   Why is it necessary to use buffered solutions in electrophoresis?

2.   Explain the principle of the staining/destaining process.

3.   What are the respective roles of the persulphate, TEMED, the glycerol and
     the bromophenol blue?

4.   What would be the advantages of isoelectric focusing in this analysis?

Instructor's Notes

1. The separations can be achieved using disc electrophoresis, but this system is more difficult for inexperienced students to use.

2. The buffer system is stable for about 4 to 6 weeks.

3. Because of the time required for destaining, students cannot assess their results during a single working session.